Universitext

Krzysztof P. Rybakowski

The Homotopy Index and Partial Differential Equations

Springer-Verlag
Berlin Heidelberg New York
London Paris Tokyo

Krzysztof P. Rybakowski
Albert-Ludwigs-Universität
Institut für Angewandte Mathematik
Hermann-Herder-Str. 10
7800 Freiburg i. Br., FRG

Mathematics Subject Classification (1980): primary 58 E05,
secondary 35 B40

ISBN 3-540-18067-2 Springer-Verlag Berlin Heidelberg New York
ISBN 0-387-18067-2 Springer-Verlag New York Berlin Heidelberg

Printing and bookbinding: Druckhaus Beltz, Hemsbach
2141/3140-543210

This book is dedicated to my parents
Maria and Reinhold Rybakowski

Introduction

The homotopy index theory was developed by Charles Conley for two-sided flows on compact spaces.
The homotopy or Conley index, which provides an algebraic-topological measure of an isolated invariant set, is defined to be the homotopy type of the quotient space N_1/N_2, where $\langle N_1,N_2 \rangle$ is a certain compact pair, called an index pair. Roughly speaking, N_1 isolates the invariant set and N_2 is the "exit ramp" of N_1.

It is shown that the index is independent of the choice of the index pair and is invariant under homotopic perturbations of the flow. Moreover, the homotopy index generalizes the Morse index of a nondegenerate critical point p with respect to a gradient flow on a compact manifold. In fact if the Morse index of p is k, then the homotopy index of the invariant set {p} is Σ^k - the homotopy type of the pointed k-dimensional unit sphere.

The homotopy invariance makes Conley's homotopy index a useful tool in global perturbation problems involving ordinary differential equations. On the other hand, when trying to apply Conley's original index to, say, parabolic partial differential equations, we face a difficulty. First of all the "flow" defined by a PDE is, in fact, only a semiflow, i.e. in general, solutions are only defined for nonnegative times. More importantly, the appropriate phase space of the equation is an infinite-dimensional Banach space, which, therefore, is not (locally) compact. Consequently, before being able to apply Conley's original index theory, one has to perform a reduction of the original problem to a finite-dimensional problem. Such a reduction imposes very stringent assumptions on the equation in question, and is, therefore, possible only in exceptional cases. Furthermore, even in case a reduction is possible, one loses in this way almost all information about the dynamics of the PDE.
In a series of papers the present author extended the homotopy index theory to certain (one-sided) semiflows on non-locally compact

metric spaces. The semiflows are required to satisfy a certain admissibility assumption which turns out to be closely related to the Palais-Smale condition in Morse theory on Banach manifolds. Admissibility is also related to the property of being an α-contraction. Thus the extension of the original Conley index is, on the one hand, analogous to the Palais-Smale generalization of the classical Morse index, and, on the other hand, it is analogous to the Leray-Schauder and Nussbaum extensions of the Brouwer mapping degree.
There is one important difference though: the definitions of the Nussbaum or Leray-Schauder indices are ultimately reduced to the definition of the Brouwer-degree, via certain finite dimensional approximations of the original mapping. No such general approximation scheme of an infinite-dimensional semiflow by finite-dimensional flows seems available, even if the admissibility assumption is satisfied. Therefore the homotopy index has to be defined directly for the infinite-dimensional semiflow in question. This requires new ideas, since Conley's arguments rest very heavily upon the compactness assumption and therefore, they cannot be relaxed as such.
The extended homotopy index theory as well as some applications are presented in the papers Rybakowski [4]-[15] and Rybakowski and Zehnder [1].
It is the purpose of this book to provide a unified approach to the theory and to some of the applications contained in the above papers. We have also tried to render the material accessible to readers with only modest knowledge of algebraic topology.

The book consists of three chapters.
In Chapter I the homotopy index theory is developed. We show the existence of a special connected simple system, called the categorial Morse index. We also define the homotopy index and prove its homotopy invariance.

Chapter II is devoted to some applications of the theory to nonlinear elliptic and parabolic equations and to the periodic boundary value problems for second-order gradient systems.

In Chapter III we prove a generalized Morse equation for a Morse decomposition of an invariant set and obtain a formula relating the Conley index to the Brouwer degree.
We also examine the connection between the admissibility assumption and the Palais-Smale condition.
Next we prove that the so-called critical groups are nothing else but the homology groups of the homotopy index.

Finally, we prove the homotopy invariance property of the categorial Morse index along paths.
The prerequisites for the understanding of the book are
(1) Knowledge of the basic concepts of topology and homology theory.
(2) Familiarity with parabolic equations as contained in the first chapters of Henry [1], together with the maximum principles.

This book originated in courses which I taught at the Universidade de São Paulo, São Carlos, Brazil, in the summer of 1983 and at the Université Catholique de Louvain, Louvain-La-Neuve, Belgium, in the spring of 1985.

I am indebted to Professor Antonio F. Izé and the Fundação de Amparo à Pesquisa do Estado de São Paulo (FAPESP) for inviting me to São Carlos.

My sincere thanks also go to Professor Jean Mawhin for his invitation to Louvain-La-Neuve and for some illuminating conversations which led to an improvement of Section 2.8.

It is a pleasure to thank Mrs. Brigitte Zakrzewski for her superb job in typing the manuscript.

Lastly, I would like to express my deep gratitude to Professor Jack Hale, who, in 1978, introduced me to infinite-dimensional dynamical systems. Without his inspiration and the discussions we had over these past years this book would not have been written.

Freiburg im Breisgau, Juni 1987 Krzysztof P. Rybakowski

Contents

Chapter I The homotopy index theory 1

1.1 Local semiflows 1

1.2 The no blow-up condition.
Convergence of semiflows 5

1.3 Isolated invariant sets and isolating blocks 6

1.4 Admissibility 13

1.5 Existence of isolating blocks 18

1.6 Homotopies and inclusion induced maps 26

1.7 Index and quasi-index pairs 29

1.8 Some special maps used in the construction of
the Morse index 33

1.9 The Categorial Morse index 39

1.10 The homotopy index and its basic properties 49

1.11 Linear semiflows. Irreducibility 57

1.12 Continuation of the homotopy index 64

Chapter II Applications to partial differential equations .. 72

2.1 Sectorial operators generated by partial
differential operators 73

2.2 Center manifolds and their approximation 76

2.3 The index product formula 78

2.4 A one-dimensional example 89

2.5 Asymptotically linear systems 94

2.6 Estimates at zero and nontrivial solution of
elliptic equations 109

2.7 Positive heteroclinic orbits of second-order
parabolic equations 117

2.8 A homotopy index continuation method and periodic
solutions of second-order gradient systems 125

Chapter III Selected topics 140

3.1 Repeller-attractor pairs and
Morse decompositions 141

3.2 Block pairs and index triples 147

3.3 A Morse equation 155

3.4 The homotopy index and Morse theory on Hilbert manifolds 163

3.5 Continuation of the categorial Morse index along paths 179

Bibliographical notes and comments 195

Bibliography 200

Index 206

Chapter I

The homotopy index theory

In this chapter we develop the concepts of the categorial Morse index and the homotopy index.
We begin by defining local semiflows and showing how they are generated by various classes of differential equations. Then we define the no-blow-up condition and the notion of convergence of semiflows.
In Section 1.3 the central concepts of an isolated invariant set, an isolating neighborhood and an isolating block are introduced.
The whole index theory as presented here rests on the concept of admissibility, introduced and discussed in Section 1.4.
Using this concept we prove (in Section 1.5) the existence of isolating blocks.
In the subsequent Sections 1.6, 1.7, 1.8 we recall certain concepts from the homotopy theory, define index and quasi-index pairs and prove a few technical results needed later on.
In Section 1.9 we define the categorial Morse index and prove that it is a connected simple system. This result enables us in Section 1.10 to define the homotopy index. We then discuss the join and the smash product of indices, compute the index of a hyperbolic equilibrium for a linear semiflow and give a criterion for the existence of heteroclinic orbits inside an invariant set (Sections 1.10 and 1.11).
Finally, in Section 1.12 we prove the homotopy invariance property of the homotopy index.

1.1 Local semiflows

In this section we introduce the concept of a local semiflow (a local semidynamical system), and discuss a few examples.

Definition 1.1
Let X be a topological space, D be an open set in $\mathbb{R}^+ \times X$ (where

$\mathbb{R}^+ = \{t \in \mathbb{R} \mid t \geq 0\}$), and $\pi: D \to X$ be a mapping. We write $x\pi t := \pi(t,x)$.

π is called a <u>local semiflow on</u> X, if the following properties are satisfied:

(1) π is continuous.

(2) For every $x \in X$ there is an ω_x, $0 < \omega_x \leq \infty$, such that $(t,x) \in D$ if and only if $0 \leq t < \omega$.

(3) $x\pi 0 = x$ for all $x \in X$.

(4) If $(t,x) \in D$ and $(s, x\pi t) \in D$, then $(t+s,x) \in D$ and $x\pi(t+s) = (x\pi t)\pi s$.

<u>Remarks</u>

If $\omega_x = \infty$ for every $x \in X$, then π is often called a <u>global semiflow</u>. If $A \subset X$ and $S \subset \mathbb{R}^+$ and if $S \times A \subset D$ then we define $A\pi S$ to be $\pi(S \times A)$. Note that we do not define $A\pi S$ if $S \times A \not\subset D$. If $A = \{x\}$ or $S = \{t\}$ we write $x\pi S := \{x\}\pi S$ and $A\pi t := A\pi\{t\}$. In most cases we will have $A = \{x\}$ and $S = [a,b]$, i.e. S is an interval.

We will now consider a few examples:

<u>Example 1.2</u> (<u>ordinary differential equations</u>)

Let Ω be an open subset of $\mathbb{R}^m$, and $f: \Omega \to \mathbb{R}^m$ be a continuous mapping. Consider the ODE

$$\dot{x} = f(x(t)) \ , \ x(0) = x_0 \ . \tag{1}$$

Assume uniqueness of solutions of (1), e.g. assume that f is locally Lipschitzian.
Then for every x_0 there is a unique noncontinuable solution $t \to x(t,x_0)$ of (1) defined on some interval $[0,\omega_x)$.

Writing $x_0\pi t := x(t,x_0)$ we obtain a local semiflow $\pi = \pi_f$ on $X = \Omega$. This is a classical elementary result from the theory of ODEs.

The solutions of (1) can also be uniquely defined for $t \in (\omega_x^*, 0]$. Therefore (1) in fact defines a local (two-sided) flow, a feature peculiar to ODEs but not to FDEs or PDEs.

<u>Example 1.3</u> (<u>retarded functional differential equations</u>)

Let $C = C([-r,0], \mathbb{R}^m)$ be equipped with the supremum-norm, $r \geq 0$, and Ω be open in C.
If $t \in \mathbb{R}$ and $x: [-r+t, t] \to \mathbb{R}^m$ is a continuous mapping, then x_t is the element of C defined as $x_t(\theta) = x(t+\theta)$, $\theta \in [-r,0]$.
Let $f: \Omega \to \mathbb{R}^m$ be a continuous mapping.

Consider the following retarded functional differential equation:

$$\dot{x}(t) = f(x_t) \qquad x_0 = \varphi. \tag{2}$$

A solution of (2) is a continuous mapping $x:[-r,A)\to\mathbb{R}^m$, $A>0$, such that $x_0=\varphi$, x is differentiable on $[0,A)$, $x_t\in\Omega$ and $\dot{x}(t)=f(x_t)$ for all $t\in[0,A)$. If f satisfies some mild assumptions, e.g. if f is Lipschitzian on compact sets in Ω, then through every $\varphi\in\Omega$ there is a unique noncontinuable solution $t\to\mathbf{x}(\varphi)(t)$ defined on some interval $[-r,\omega_\varphi)$. Writing $\varphi\pi t=x_t(\varphi)$ for $t\in[0,\omega_\varphi)$ we obtain a local semiflow $\pi=\pi_f$ on $X=\Omega$. Taking $r=0$, we get $C\simeq\mathbb{R}^m$ and so (1) is a special case of (2).

<u>Example 1.4</u> (<u>neutral functional differential equations</u>)

Let C, Ω and f be as in Example 1.3. Moreover, let $D:C\to\mathbb{R}^m$ be a bounded linear mapping. By Riesz's theorem, there is a mapping $\eta:[-r,0]\to L(\mathbb{R}^m,\mathbb{R}^m)$ of bounded variation such that

$$D(\varphi) = \int_{-r}^{0} d_\theta\eta\cdot\varphi(\theta) \qquad \text{for all } \varphi\in C .$$

D is called a <u>stable difference operator</u> if (1) and (2) below hold:

(1) D is atomic at zero, i.e. $\eta(0)-\lim_{\theta\to 0^-}\eta(\theta)$ is nonsingular.

(2) There are constants $a,b>0$ such that for every $h\in C(\mathbb{R}^+,\mathbb{R}^m)$ and every solution $y:[-r,\infty)\to\mathbb{R}^m$ of $Dy_t=h(t)$, $t\geq 0$, it follows that

$$\|y_t\| \leq be^{-at}\|y_0\| + b\sup_{0\leq u\leq t}\|h(u)\| .$$

Let D be a stable difference operator and consider the neutral functional differential equation

$$\frac{dD(x_t)}{dt} = f(x_t) \qquad x_t = \varphi . \tag{3}$$

A solution of (3) is a continuous mapping $x:[-r,A)\to\mathbb{R}^m$, $A>0$, with $x_0=\varphi$ and $x_t\in\Omega$ for all $t\in[0,A)$, and such that $t\to D(x_t)$ is differentiable on $[0,A)$ and

$$\frac{dD(x_t)}{dt} = f(x_t) \quad \text{for } t\in[0,A) .$$

If f is Lipschitzian on compact sets in Ω, then for every $\varphi\in\Omega$

there is a unique solution $t \to x(\varphi)(t)$ of (3) defined on some maximal interval $[-r,\omega_\varphi)$. Writing $\varphi\pi t = x_t(\varphi)$, $t \in [0,\omega_\varphi)$, we obtain a local semiflow $\pi = \pi_f$ on $X = \Omega$. Taking $D\varphi := \varphi(0)$ for $\varphi \in C$, we see that equation (2) is a special case of equation (3).
The proofs of all these assertions can be found in Hale [1].

Example 1.5 (semilinear parabolic equations):

We now discuss, in an abstract setting, semilinear parabolic partial differential equations. For details, we refer the reader to Dan Henry's monograph, Henry [1].

Let X be a Banach space, and $A: D(A) \to X$ be a closed, densely defined, linear operator in X. A is called sectorial, if there are constants Φ, M, a, $0<\Phi<\pi/2$, $M>1$, $a \in \mathbb{R}$ such that the sector $S_{\Phi,a} = \{\lambda \in \mathbb{C} \mid \lambda \neq a, \Phi < |\arg(\lambda - a)| \leq \pi\}$ is contained in $\rho(A)$, the resolvent set of A, and $\|(\lambda - A)^{-1}\| \leq M/|\lambda - a|$ for all $\lambda \in S_{\Phi,a}$. If A is sectorial, then there is a $k \geq 0$ such that $\operatorname{Re}\sigma(A+kI) > 0$. For $A_1 := A + kI$ and $0<\alpha<1$ define

$$A_1^{-\alpha} = \frac{\sin \pi\alpha}{\pi} \int_0^\infty \lambda^{-\alpha}(\lambda + A_1)^{-1} d\lambda .$$

$A_1^{-\alpha}$ is bounded and injective. Let X^α be the range of $A_1^{-\alpha}$, $X^0 = X$, $X^1 = D(A)$. Let $A_1^\alpha: X^\alpha \to X$ be the inverse of $A_1^{-\alpha}$, $A_1^0 = Id_X$, $A_1^1 = A_1$. X^α is dense in X. Define the norm $\| \;\|_\alpha$ on X by $\|u\|_\alpha = \|A_1 u\|$, where $\| \;\|$ is the norm of X. X^α does not depend on the choice of k, and different choices of k yield equivalent norms on X^α. X^α is a Banach space under $\|.\|_\alpha$, called the α-th fractional power space of X (rel. to A).

Suppose $0 \leq \alpha < 1$, U is open in X^α and $f: U \to X$ is a locally Lipschitz continuous mapping. Consider the equation (S_f):

$$\frac{du}{dt} + Au = f(u) \qquad (S_f) .$$

Let $u_0 \in U$. By a solution of (S_f) on $(0,t_0)$ through u_0 we mean a continuous mapping $u: [0,t_0) \to X$, such that $u(0) = u_0$, u is differentiable on $(0,t_0)$, $u(t) \in D(A)$ for $t \in (0,t_0)$, $t \to f(u(t))$ is locally Hölder continuous, $\int_0^a \|f(u(t))\| dt < \infty$ for some $a>0$, and (S_f) holds for $t \in (0,t_0)$.
In this definition, "$t \to g(t)$ is locally Hölder-continuous" means that for every t there exists a neighborhood W of t and $L, \theta > 0$ such that

$$\|g(t_1) - g(t_2)\| \leq L|t_1 - t_2|^\theta$$

for $t_1, t_2 \in W$.

Theorem 1.6 (see Henry [1], Theorems 3.3.3 and 3.4.1).

Let A be sectorial on X, and let $f: U \to X$ be locally Lipschitzian where U is an open subset of X^α for some $0 \leq \alpha < 1$. Then for every $u_0 \in U$ there exists a maximal $0 < \omega_{u_0} \leq \infty$ and a unique solution $t \to u(t, u_0)$ of (S_f) through u_0, defined on $[0, \omega_{u_0})$. Writing $u_0 \pi_f t = u(t, u_0)$, we obtain a local semiflow π_f on U.

1.2 The no-blow-up condition. Convergence of semiflows

If π is a local semiflow on X and $x \in X$ with $\omega_x < \infty$, then this usually means that the solution $x\pi t$ 'blows up' in some sense. One is naturally interested in ruling out such cases:

Definition 2.1

Let N be an arbitrary subset of X. We say that the local semiflow *does not explode* in N, if whenever $x \in X$ and $x\pi[0, \omega_x) \subset N$, then $\omega_x = \infty$.

In these notes we will use different local semiflows on the same metric space X. We may e.g. think of the nonlinearity f in (S_f) as containing a parameter λ and see how π_f depends on this parameter. Under natural assumptions this dependence will be continuous.

This leads to the following convergence concept:

Definition 2.2

Let π and π_n, $n \in \mathbb{N}$, be local semiflows on X. We say that the sequence $\{\pi_n\}$ *converges to* π (we write $\pi_n \to \pi$ as $n \to \infty$) if for every $x \in X$, $t \in \mathbb{R}^+$ and every two sequences $\{x_n\}$ and $\{t_n\} \subset \mathbb{R}^+$, whenever $x_n \to x$ and $t_n \to t$ as $n \to \infty$ and $x\pi t$ is defined, then $x_n \pi_n t_n$ is defined for all n sufficiently large, and $x_n \pi_n t_n \to x\pi t$ as $n \to \infty$. If $\pi_n \to \pi, n \to \infty$, then it is easily proved that $\pi_{n_m} \to \pi, m \to \infty$ for every sequence $\{\pi_{n_m}\}$ of $\{\pi_n\}$.

Theorem 2.3

Let $D: C \to \mathbb{R}^m$ be a stable difference operator, and $\Omega \subset C$ be open. Let $f: \Omega \to \mathbb{R}^m$ and $f_n: \Omega \to \mathbb{R}^m$, $n \in \mathbb{N}$ be locally Lipschitzian mappings. Then the following properties hold:

(1) If $N \subset \Omega$, N is bounded and closed in C and $f[N]$ is bounded, then the local semiflow π generated by

$$\frac{d}{dt} D(x_t) = f(x_t)$$

does not explode in N.

(2) If $f_n(x) \to f_0(x)$ as $n \to \infty$ uniformly on compact subsets of Ω, then $\pi_n \to \pi_0$ as $n \to \infty$, where for $n \geq 0$, π_n is the local semiflow generated by solutions of

$$\frac{dD(x_t)}{dt} = f_n(x_t) \ .$$

The proof of Theorem 2.3 follows from results in Hale [1] and is omitted. Specializing to $D\varphi \equiv \varphi(0)$ we get a similar statement for ODEs and RFDEs.

We also obtain an analogous statement for parabolic PDEs:

Theorem 2.4

Let A be sectorial on X and let U be an open subset of X^α for some $0 \leq \alpha < 1$. Then the following properties hold:

(1) If $N \subset U$, N is closed in X^α and N is bounded with respect to the norm of $X^0 = X$, and if $f: U \to X$ is locally Lipschitzian with $f[N]$ bounded, then π_f does not explode in N.

(2) Let $f: U \to X$ and $f_n: U \to X$, $n \in \mathbb{N}$, be locally Lipschitzian mappings. Suppose that $f_n(u) \to f(u)$ as $n \to \infty$ locally uniformly in u, i.e. for every $u_0 \in U$ there is a neighborhood $V \subset U$ of u_0 such that $f_n(u) \to f(u)$ as $n \to \infty$ uniformly on V. Then $\pi_{f_n} \to \pi_f$ as $n \to \infty$.

Proof :

Part 1 of the Theorem 2.4 is essentially Theorem 3.3.4 in Henry [1], whose proof can be used almost verbatim in our case.
Part 2 is a restatement of Theorem 3.4.1 in Henry [1] on continuous dependence of solutions of (S_f) on time t, initial values and parameters.

1.3 Isolated invariant sets and isolating blocks

Let X be a metric space and π be a local semiflow on X. Let J be an interval in $\mathbb{R}$ and $\sigma: J \to X$ be a mapping. σ is called a solution (of π) if for all $t \in J$, $s \in \mathbb{R}^+$ for which $t+s \in J$, it follows that $\sigma(t)\pi s$ is defined and $\sigma(t)\pi s = \sigma(t+s)$. If $0 \in J$ and $\sigma(0) = x$, then we say that σ is a solution through x. If $J = (-\infty, \infty)$, then σ is called a full solution.

If σ is a solution defined on $\mathbb{R}^+ = [0, \infty)$ (resp. on $\mathbb{R}^- = (-\infty, 0]$) then by its ω-limit set $\omega(\sigma)$ (resp. α-limit set $\omega^*(\sigma)$, also denoted by $\alpha(\sigma)$) we mean the set of all $y \in X$ for which there is a sequence $t_n \to +\infty$ (resp. $t_n \to -\infty$) with $\sigma(t_n) \to y$ as $n \to \infty$. $\omega(\sigma)$ or $\omega^*(\sigma)$ may be empty.

Note that if σ is defined on $\mathbb{R}^+$ then $\sigma(t)=\sigma(0)\pi t$ for all t, and $\omega(\sigma)$ depends only on $x_0=\sigma(0)$. In this case we often write $\omega(x_0)$. For two-sided flows one can also write $\omega^*(x_0)$ instead of $\omega^*(\sigma)$. However, in our more general case of a local semiflow $\omega^*(\sigma)$ depends, in general, on the whole solution.

Suppose now that Y is a subset of X, and set:

$$A_\pi^+(Y) = A^+(Y) = \{x \in X \mid x\pi[0,\omega_x) \subset Y\}$$

$$A_\pi^-(Y) = A^-(Y) = \{x \in X \mid \text{there is a solution } \sigma:(0,-\infty]\to X \text{ through } x \text{ with } \sigma(-\infty,0]\subset Y\}.$$

$$A_\pi(Y) = A(Y) = A_\pi^+(Y)\cap A_\pi^-(Y).$$

Y is called <u>positively invariant</u> if $Y=A_\pi^+(Y)$,
Y is called <u>negatively invariant</u> if $Y=A_\pi^-(Y)$,
Y is called <u>invariant</u> if $Y=A_\pi(Y)$.

(Note that all these concepts are relative to π).

In particular, if $\omega_x=\infty$ for every $x \in Y$, then Y is invariant iff for every $x \in Y$ there exists a full solution σ through x for which $\sigma(\mathbb{R})\subset Y$.

If σ is defined on $\mathbb{R}^+$ (resp. on $\mathbb{R}^-$) and $\sigma[\mathbb{R}^+]$ (resp. $\sigma[\mathbb{R}^-]$) is relatively compact, then it is well-known that $\omega(\sigma)$ (resp. $\alpha(\sigma)$) is a nonempty, compact, connected and invariant subset of X (see e.g. Bhatja and Hâjek [1]).

For a general subset Y of X, $A^+(Y)$ (resp. $A^-(Y)$, resp. $A(Y)$) is easily seen to be the largest positively invariant (resp. negatively invariant, resp. invariant) subset of Y. $A^+(Y)$ (resp. $A^-(Y)$) is often called the <u>stable</u> (resp. <u>unstable) manifold</u> of $K=A(Y)$, <u>relative</u> to Y.

To illustrate these concepts with an example, suppose that f is an RFDE on $\mathbb{R}^m$ of class C^1 and 0 is a hyperbolic equilibrium of f (cf. Hale [1], Chapter 10). Then the well-known saddle-point property implies that there is a direct sum decomposition $C=U\oplus S$ and a closed neighborhood Y of 0 such that $K=\{0\}$ is the largest invariant set in Y, i.e., $\{0\}=A(Y)$. Moreover, the sets $A^+(Y)$ and $A^-(Y)$ are tangent to S, resp. to U at zero. There is a small ball $B_\delta\subset Y$ such that $A^+(Y)\cap B_\delta$ (resp. $A^-(Y)\cap B_\delta$) is diffeomorphic to $S\cap B_\delta$ (resp. $U\cap B_\delta$). Finally, the ω-limit set of every solution starting in $A^+(Y)$ (resp. the α-limit set of every solution defined on $(-\infty,0]$ and remaining in $A^-(Y)$) is equal to $\{0\}$. Therefore, the qualitative picture near the equilibrium looks as in Fig. 1.

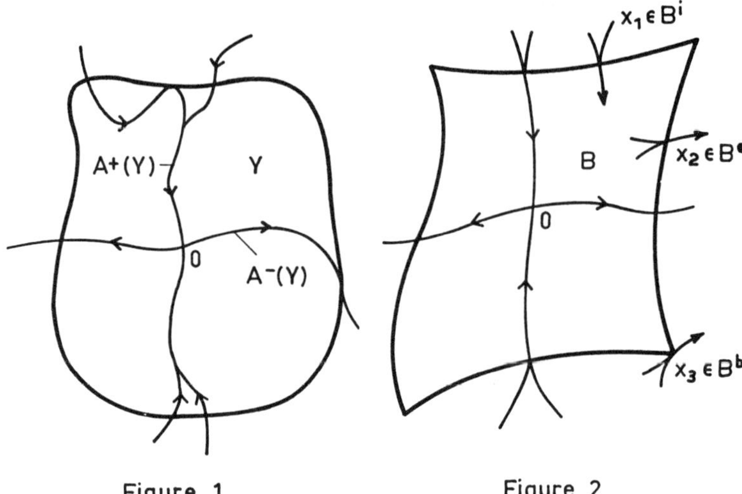

Figure 1 Figure 2

The set K={0} has the important property of being isolated by Y. More generally, we have the following fundamental

Definition 3.1

If K is a closed invariant set (relative to π) and if there is a closed neighborhood N of K (i.e. N is closed and $K\subset \mathrm{Int}\, N$) such that K=A(N), then K is called an _isolated invariant set (relative to π)_. Conversely, if N is a closed subset of X such that K=A(N) is closed and $K\subset \mathrm{Int}\, N$, then N is called an _isolating neighborhood (of K) (relative to π)_.

In the situation of Fig. 1, K={0} is, in fact, an isolated invariant set and Y is an isolating neighborhood of K. Let us analyze this example a little further: The isolating neighborhood Y is rather arbitrary, i.e., its boundary ∂Y is unrelated in any way to the semiflow π. However, Fig. 2 suggests that one should be able to choose the set Y in such a way that ∂Y is "transversal" to π, i.e., such that orbits of solutions of π cross Y in one or the other time direction. In fact, this is, for example, the case for ODEs, where such special sets, called isolating blocks, are used in connection with the famous Ważewski principle. The transversality of ∂Y with respect to π implies that every point $x \in \partial Y$ is of one of the following three types: it is either a strict egress, or a strict ingress or a bounce-off point.

Let us define those three concepts for an arbitrary local semiflow π.

<u>Definition 3.2</u>

Let $B \subset X$ be a closed set and $x \in \partial B$ a boundary point. Then x is called a <u>strict egress</u> (resp. <u>strict ingress</u>, resp. <u>bounce-off</u>) point of B, if for every solution $\sigma:[-\delta_1,\delta_2] \to X$ through $x=\sigma(0)$, with $\delta_1 \geq 0$ and $\delta_2>0$, the following properties hold:

(1) There is an ε_2, $0<\varepsilon_2 \leq \delta_2$ such that $\sigma(t) \notin B$ (resp. $\sigma(t) \in \text{Int } B$, resp. $\sigma(t) \notin B$), for $0<t \leq \varepsilon_2$.

(2) If $\delta_1>0$ then for some ε_1 with $0<\varepsilon_1<\delta_1$: $\sigma(t) \in \text{Int } B$ (resp. $\sigma(t) \notin B$, resp. $\sigma(t) \notin B$), for $-\varepsilon_1 \leq t<0$.

By B^e (resp. B^i, resp. B^b) we denote the set of all strict egress (resp. strict ingress, resp. bounce-off) points of the closed set B. We finally set $B^+=B^i \cup B^b$ and $B^-=B^e \cup B^b$. We then have the following:

<u>Definition 3.3</u> (<u>Isolating block</u>)

A closed set $B \subset X$ is called an <u>isolating block</u>, if

(i) $\partial B = B^e \cup B^i \cup B^b$.

(ii) B^- is closed.

Note that for general semiflows, $B^e \cap B^b$ may be nonempty , and consist of points $x \in B^e$ for which there is no solution defined for any negative times.

Given an isolated invariant set K, is it possible to construct an isolating neighborhood of K in the form of an isolating block?

Fig. 2 suggests that this should be true for the special case of a hyperbolic equilibrium, but we shall try to give a general answer. However, we shall first prove the following simple result about isolated invariant sets which has recently been applied to prove persistence of populations modeled by ordinary differential equations or reaction-diffusion systems.

<u>Theorem 3.4</u>

<u>Let K be an isolated invariant set and N be an isolated neighborhood of K, such that π does not explode in N. Let $y \in X$ be such that $\omega_y=\infty$ and $y\pi[0,\infty)$ is relatively compact. If $\omega(y) \cap K \neq \emptyset$ and $\omega(y) \setminus K \neq \emptyset$, then</u>

$A^+(N)\cap\partial N\cap\omega(y)\neq\emptyset$ and $A^-(N)\cap\partial N\cap\omega(y)\neq\emptyset$.

Remark: Theorem 3.4 roughly says that if a solution gets infinitely close to K without remaining close, then this solution must get infinitely close to $A^+(N)\setminus K$ and $A^-(N)\setminus K$.

Proof:

Let $z_0\in\omega(y)\cap K$. Then there exists a sequence $\{t_n\}_{n\in\mathbb{N}}\subset\mathbb{R}^+$ such that $t_n\to\infty$, $y\pi t_n\to z_0$, as $n\to\infty$. W.l.o.g. $y\pi t_n\in \operatorname{Int} N$, for $n\in\mathbb{N}$.
Since K is maximal invariant in N and $\omega(y)$ is invariant and $\omega(y)\not\subset K$, it follows that $\omega(y)\not\subset N$. Choose $z_0'\in\omega(y)\setminus N$. Then there is a sequence $\{t_n'\}_{n\in\mathbb{N}}\subset\mathbb{R}^+$ with $t_n'\to\infty$ and $y\pi t_n'\to z_0'$ as $n\to\infty$. Since N is closed by definition we can assume that $y\pi t_n'\notin N$ for $n\in\mathbb{N}$. By taking subsequences and renumbering them, if necessary, we can assume that

$$t_{2n}' < t_n < t_{2n+1}' \quad , \quad \text{for all } n\in\mathbb{N} .$$

Define

$$s_{2n} = \inf\,\{t \mid t_{2n}'<t<t_n \quad \text{and } y\pi[t,t_n]\subset N\}$$
$$s_{2n+1} = \sup\,\{t \mid t_n<t<t_{2n+1}' \quad \text{and } y\pi[t_n,t]\subset N\}.$$

Then, obviously

$$t_{2n}' < s_{2n} < t_n < s_{2n+1} < t_{2n+1}' \quad ,$$
$$y\pi[s_{2n},s_{2n+1}]\subset N,\ y\pi s_{2n}\in\partial N,\ y\pi s_{2n+1}\in\partial N,\ \text{for } n\in\mathbb{N} .$$

Again we may assume w.l.o.g. that

$$y\pi s_{2n}\to z_1 \quad \text{and } y\pi s_{2n+1}\to z_2 \quad \text{as } n\to\infty$$

for some $z_1,z_2\in X$. It follows that $z_1,z_2\in\omega(y)\cap\partial N$.
We will show that $z_1\in A^+(N)$ and $z_2\in A^-(N)$, thereby completing the proof. First note that since π does not explode in N, $K\subset \operatorname{Int} N$ and K is invariant, we have $z_0\pi[0,\infty)$ is defined and $z_0\pi[0,\infty)\subset \operatorname{Int} N$. A simple continuity argument implies that $s_{2n+1}-t_n\to\infty$ as $n\to\infty$. Let $t\geq 0$ be arbitrary. Then there is an n_0 such that for all $n\geq n_0$, $s_{2n}+t<s_{2n+1}$.
Thus $z_1\pi t=\lim_{n\to\infty} y\pi(s_{2n}+t)\in N$ for all $t\geq 0$ for which $z_1\pi t$ is defined.

Since π does not explode in N it follows that $z_1\pi[0,\infty)$ is defined and $z_1\pi[0,\infty)\subset N$, i.e. $z_1 \in A^+(N)$.

Now note that the sequence $\{y\pi(s_{2n+1}-1) \mid n\geq n_0\}$ is defined for n_0 large and contains a subsequence converging to some $a_{-1} \in X$. Define $\sigma:[-1,0]\to X$ by $\sigma(-1+t)=a_{-1}\pi t$ for $t \in [0,1]$. Then σ is a well-defined solution of $\pi, \sigma(0)=z_2$ and σ lies in N.

Continuing this type of argument we obtain a solution σ as in the definition of $A^-(N)$ with $\sigma(0)=z_2$. Hence $z_2 \in A^-(N)$ and the proof is complete.

Corollary 3.5

Under the assumptions ot Theorem 3.4 the sets $(A^+(N)\setminus K)\cap\omega(y)$ and $(A^-(N)\setminus K)\cap\omega(y)$ are infinite.

Proof:

Let V=Int N. Then $K\subset V\subset N$. Since X is a normal space, there are open sets V_1, W_1 such that $V_1\cap W_1=\emptyset$ and $K\subset V_1$, $X\setminus W_1\subset V$. Hence $K\subset V_1\subset X\setminus W_1\subset V$. Continuing this argument we obtain two sequences $\{V_n\}$ and $\{W_n\}$ of open sets such that for all n,

$$K\subset V_{n+1}\subset X\setminus W_{n+1}\subset V_n\subset X\setminus W_n\subset V\ .$$

Applying Theorem 3.4 to $N_n:=X\setminus W_n$ and noting that $A^+(N_n)\subset A^+(N)$ and $A^-(N_n)\subset A^-(N)$ obtain sequences $\{u_n\}$, $\{v_n\}$ with $u_n \in A^+(N)\cap\partial N_n\cap\omega(y)$, $v_n \in A^-(N)\cap\partial N_n\cap\omega(y)$.

Obviously $u_n\neq u_m$ and $v_n\neq v_m$ for $n\neq m$, proving the corollary.

We shall end this section by proving an important property of isolating blocks.

Recall that if B is a topological space and $A\subset B$ is a subspace, then the inclusion $A\subset B$ is called a cofibration if it has the homotopy extension property with respect to any space Y, i.e. if given continuous maps $g:B\to Y$ and $G:A\times[0,1]\to Y$ such that $g(x)=G(x,0)$ for $x \in A$, there is a continuous map $F:B\times[0,1]\to Y$ with $F(x,0)=g(x)$ for $x \in X$ and $F|A\times[0,1]=G$. Using Exercise E6 in Chapter I of Spanier [1] and simple properties of metric spaces one can easily prove

Proposition 3.6

If B is a metric space and $A\subset B$ is closed, then $A\subset B$ is a cofibration if and only if there is an open neighborhood U of A in B and a conti-

nuous map $H:U\times[0,1]\to B$, *such that* $H(x,0)=x$, $H(a,t)=a$ *and* $H(x,1)\in A$ *for* $x\in U$, $a\in A$ *and* $t\in[0,1]$.

Now we have

Theorem 3.7

If B *is an isolating block (rel. to* π*) and* π *does not explode in* B, *then the inclusion* $B^-\subset B$ *is a cofibration.*

To prove this theorem we need the following lemma, which is of interest in its own right.

Lemma 3.8

Under the assumptions of Theorem 3.7 *define*
$s_B:B\to\mathbb{R}^+\cup\{\infty\}$ by $s_B(x)=\sup\{t<\omega_x \mid x\pi[0,t]\subset B\}$.
Then s_B *is continuous.*

Proof:

Let $x_n\in B$, $x\in B$, and $x_n\to x$. Assume first $s_B(x)<\infty$. Since π does not explode in B, it follows that $s_B(x)<\omega_x$ and $x\pi s_B(x)\in\partial B$. Definition 3.3 immediately implies $x\pi s_B(x)\in B^-$. Therefore, for every small $\varepsilon>0$, $x\pi(s_B(x)+\varepsilon)\in X\setminus B$, and by continuity $x_n\pi(s_B(x)+\varepsilon)$ is defined and $\in X\setminus B$ for n large, so $s_B(x_n)<s_B(x)+\varepsilon$.

Now let $0<s_B(x)\leq\infty$. Since B is an isolating block, it follows that $x\pi[0,s_B(x))$ is defined and $x\pi(0,s_B(x))\subset\text{Int } B$. Therefore, for arbitrary ε,T with $0<\varepsilon<T<s_B(x)$, there is an $n(\varepsilon,T)$ such that $x_n\pi[\varepsilon,T]$ is defined and $\subset\text{Int } B$ for $n\geq n(\varepsilon,T)$. If $s_B(x_n)$ does not converge to $s_B(x)$ then, by what we have proved so far, $s_B(x_n')\leq\varepsilon$ for a subsequence $\{x_n'\}$ of $\{x_n\}$. Thus, w.l.o.g., $s_B(x_n')\leq\varepsilon_n$ for a sequence $\varepsilon_n\to 0$, as $n\to\infty$.

Therefore $s_B(x_n')\to 0$, and therefore there is a sequence $\delta_n\to 0$ such that $x_n'\pi\delta_n\notin B$. Since solutions can leave B only through B^-, there is another sequence $0\leq\varepsilon_n<\delta_n$ satisfying $x_n'\pi\varepsilon_n\in B^-$. It follows that $x\in B^-$, i.e. $s_B(x)=0$, a contradiction, proving the lemma.

Proof of Theorem 3.7:

Let $U:=B\setminus A^+(B)$. Since π does not explode in B, it is easily proved that $A^+(B)$ is closed (in X) and so U is open in B. Moreover, $B^-\subset U$.

Define $H:U\times[0,1]\to B$ as $H(x,t)=x\pi(t\cdot s_B(x))$.

H is well-defined and continuous by Lemma 3.8.
Obviously $H(x,0)=x$ for $x\in U$,

$$H(a,t) = a\pi(t\cdot s_B(a)) = a\pi(t\cdot 0) = a \quad \text{for } a\in B^-$$

and

$$H(x,1) = x\pi s_B(x) \in B^- \quad \text{for } x \in U$$

by the definition of s_B. Therefore Proposition 3.6 implies the result.

Remark: Theorem 3.7 will turn out to be very useful in the applications of the homotopy index.

1.4 Admissibility

The homotopy index theory rests on the following concept:

Definition 4.1 (Admissibility)

Let N be a closed subset of X, and let π_n, $n \in \mathbb{N}$ be a sequence of semiflows on X. Then N is called $\{\pi_n\}$-admissible if the following holds:

If $\{x_n\} \subset X$ and $\{t_n\} \subset \mathbb{R}^+$ are two arbitrary sequences such that $t_n \to \infty$ as $n \to \infty$ and $x_n \pi [0,t_n] \subset N$ for all $n \in \mathbb{N}$, then the sequence of endpoints $\{x_n \pi t_n \mid n \in \mathbb{N}\}$ has a convergent subsequence. N is called strongly $\{\pi_n\}$-admissible if N is $\{\pi_n\}$-admissible and if π_n does not explode in N for every $n \in \mathbb{N}$. If $\pi_n \equiv \pi$ for all n we also say that N is π-admissible (resp. strongly π-admissible, as the case may be).

If π is clear from the context, we often simply say that N is admissible or strongly admissible.

For ODEs in finite dimensions, admissibility is trivial: if $\Omega \subset \mathbb{R}^m$ is open, $N \subset \Omega$ is compact and $f_n : \Omega \to \mathbb{R}^m$, $n \in \mathbb{N}$, are locally Lipschitzian, then N is obviously $\{\pi_n\}$-admissible, where π_n is the local semiflow generated by $\dot{x} = f_n(x)$. This is how Conley's homotopy index for ODEs becomes a special case of the more general theory presented here. In fact, admissibility is a condition which is quite naturally encountered in FDEs or parabolic PDEs:

Theorem 4.2

Let $D: C \to \mathbb{R}^m$ be a stable difference operator and $\Omega \subset C$ be open. Let $f_n : \Omega \to \mathbb{R}^m$ be a sequence of locally Lipschitzian mappings and $N \subset \Omega$ be a set which is closed and bounded in C and $\|f_n(N)\| \leq M < \infty$ for all n (i.e. the f_n's are all bounded on N with the same bound).
Then N is $\{\pi_n\}$-admissible, where π_n is the local semiflow generated by the solutions of

$$\frac{d}{dt} D(x_t) = f_n(x_t) \ .$$

We will prove this theorem for a special case of RFDEs, i.e. where $D\varphi=\varphi(0)$, referring the reader to Rybakowski [4] for the general case.

Let $\{\varphi_n\}$, $\{t_n\}$ be such that $\varphi_n\pi_n[0,t_n]\subset N$ and $t_n\to\infty$. Then, w.l.o.g., $t_n\geq r$. We show that $\{\varphi_n\pi_n t_n\}$ is a precompact sequence. We use the Arzelà-Ascoli Theorem. $\{\varphi_n\pi_n t_n\}$ is bounded in C, since N is bounded. To prove that $\{\varphi_n\pi_n t_n\}$ is equicontinuous, let $\theta,\theta'\in[-r,0]$ be arbitrary, say $\theta'<\theta$.

Then

$$\|(\varphi_n\pi_n t_n)(\theta)-(\varphi_n\pi_n t_n)(\theta')\| = \|\int_{t_n+\theta'}^{t_n+\theta} f_n(\varphi_n\pi_n s)ds\| \leq (\theta-\theta')\cdot M$$

where M is a bound for $\|f_n(\varphi)\|$, $\varphi\in N$.

This proves that $\{\varphi_n\pi_n t_n\}$ is equicontinuous. Consequently $\{\varphi_n\pi_n t_n\}$ is precompact in C, so this sequence contains a convergent subsequence.

It follows that N is, indeed, $\{\pi_n\}$-admissible.
Note that in the RFDE case, $t_n\geq r$ is enough to have precompactness of $\{\varphi_n\pi_n t_n \mid n\geq 1\}$. The situation is different for general NFDEs where we must assume that $t_n\to\infty$, in general (cf. Rybakowski [4]).

We shall now see how admissibility enters semilinear parabolic equations. We need the following

Theorem 4.3

Let A be sectorial in X and let U be open in X^α, $0\leq\alpha<1$. Let N be a bounded subset of X (bounded in X!). Let $L,\varepsilon>0$ and $0\leq\beta<1$ be arbitrary. Then there is a $b=b(N,L,\varepsilon,\beta)>0$ such that whenever $f:U\to X$ is a locally Lipschitzian mapping and $t\to u(t)$ is a solution of (S_f) on some interval $[0,t_0)$ such that $u(t)\in N$ and $\|f(u(t))\|_X<L$ for all $t\in[0,t_0)$, then $\|u(t)\|_\beta<b$ for $t\in[\varepsilon,t_0)$.

In particular, if A has compact resolvent, then there is a compact set $C\subset X^\alpha$, $C=C(N,L,\varepsilon)$ such that if f and u are as above, then $u(t)\in C$ for $t\in[\varepsilon,t_0)$.

Proof:

There is a $k>0$ such that $\operatorname{re}\sigma(A+kI)>\delta>0$. W.l.o.g. we can assume $\|u\|_\alpha=\|A_1^\alpha u\|$ where $A_1=A+kI$. Set $f_1(u)=f(u)+ku$. Then, by the variation-of constants formula we obtain

$$u(t) = e^{-A_1 t}u(0)+\int_0^t e^{-A_1(t-s)}f_1(u(s))ds.$$

Hence for $t>0$

$$\|u(t)\|_\beta \le \|A_1^\beta e^{-A_1 t}\|\cdot\|u(0)\|_X + \int_0^t \|A_1^\beta \cdot e^{-A_1(t-s)}\|\cdot\|f_1(u(s))\|_X ds \le$$

$$\le C_\beta t^{-\beta}\cdot e^{-\delta t} L_1 + \int_0^t C_\beta \cdot (t-s)^{-\beta} e^{-\delta(t-s)} (L+kL_1) ds \ .$$

Here $L_1 = \sup\{\|u\|_X \mid u \in N\}$, and $C_\beta > 0$ is a constant, depending on β only. (We have used Theorem 1.4.3 in Henry [1].) Hence for $t \ge \varepsilon$ we have

$$\|u(t)\|_\beta \le C_\beta L_1 \varepsilon^{-\beta} e^{-\delta\varepsilon} + C_\beta (L+kL_1) \int_0^t s^{-\beta} e^{-\delta s} ds \ .$$

However,

$$\int_0^t s^{-\beta} e^{-\delta s} ds = \int_0^\varepsilon s^{-\beta} e^{-\delta s} ds + \int_\varepsilon^t s^{-\beta} e^{-\delta s} ds \le$$

$$\le \int_0^\varepsilon s^{-\beta} ds + \varepsilon^{-\beta} \int_\varepsilon^\infty e^{-\delta s} ds = M(\varepsilon) < \infty.$$

Hence $\|u(t)\|_\beta \le b$ for $t \in [\varepsilon, t_0)$, where

$$b = C_\beta L_1 \varepsilon^{-\beta} e^{-\delta\varepsilon} + C_\beta (L+kL_1) \cdot M(\varepsilon) \ .$$

This proves the first part of the theorem. If A has compact resolvent then choose $1 > \beta > \alpha$, and $b = b(N,L,\varepsilon,\beta)$ as in the first part.

Let $C = \{u \in X^\beta \mid \|u\|_\beta \le b\}$. Then C is compact in the topology of X^α.

This proves the second part of the theorem.

As a corollary, we obtain

Theorem 4.4

Let A be sectorial on X with compact resolvent and U be open in X^α for some $0 \le \alpha < 1$.

Let $N \subset U$ be closed in X^α, bounded in X, and suppose that $\{f_n : U \to X \mid n \in \mathbb{N}\}$ is a sequence of locally Lipschitzian mappings bounded on N by a common bound (i.e. $\|f_n(u)\|_X \le L$ for some $L < \infty$, all $n \in \mathbb{N}$ and all $u \in N$). Then N is strongly $\{\pi_{f_n}\}$-admissible.

Proof:

Theorem 2.4 implies that π_{f_n} does not explode in N for all $n \in \mathbb{N}$. Let $u_n \in N$, $t_n \to \infty$ and $u_n \pi_{f_n} [0, t_n] \subset N$ for all n.

W.l.o.g $t_n \geq \varepsilon := 1$. By Theorem 4.3 there is a compact set $C=C(N,L,\varepsilon)$ in X^α such that

$$u_n \pi_{f_n} [\varepsilon, t_n] \subset C \qquad \text{for all } n.$$

This implies, in particular, that $\{u_n \pi_{f_n} t_n\}$ is precompact in X^α. Hence N is $\{\pi_{f_n}\}$-admissible.

Remark:

The hypothesis that A have compact resolvent in Theorem 4.4 can be relaxed (see Theorem 4.4 in Chapter III below).

The following result collects some basic implications of our definitions.

Theorem 4.5

Let $N \subset X$ be closed and π, π_n, $n \in \mathbb{N}$, be local semiflows on X. Assume that $\pi_n \to \pi$ as $n \to \infty$, and that π does not explode in N.

Then the following properties are satisfied:

1. For every $x \in X$ and all sequences $x_n \in X$, $t_n \in \mathbb{R}^+$, $n \in \mathbb{N}$, with $x_n \to x$ as $n \to \infty$, and $x_n \pi_n [0,t_n]$ defined and $\subset N$ for all $n \in \mathbb{N}$, it follows:

1.1 If $t_n \to \infty$, then $x\pi t$ is defined for all $t \geq 0$ and $x \in A_\pi^+(N)$.
1.2 If $t_n \to t_0 < \infty$, then $x\pi[0,t_0]$ is defined and $x\pi[0,t_0] \subset N$.

2. Assume that N is $\{\pi_{n_m}\}$-admissible for every subsequence $\{\pi_{n_m}\}$ of $\{\pi_n\}$.

Then:

2.1 If $x_n \in X$ and $t_n \in \mathbb{R}^+$, $n \in \mathbb{N}$, are such that $x_n \pi_n [0,t_n]$ is defined and $\subset N$ for all n and $t_n \to \infty$ as $n \to \infty$, then every limit point of $\{x_n \pi_n t_n\}$ belongs to $A^-(N)$.
2.2 If W is any subset of N with $A_\pi(N) \subset \operatorname{Int} W$ and if π_n does not explode in N for all $n \in \mathbb{N}$, then $A_{\pi_n}(N) \subset \operatorname{Int} W$ for all n large enough.
2.3 If $\pi_n \equiv \pi$ for all n, then $A_\pi(N)$ and $A_\pi^-(N)$ are compact.

Proof:

1. Suppose $x_n \pi_n [0,t_n] \subset N$ and $x_n \to x$. Let $0 < \omega \leq \infty$ be such that $x\pi t$ is defined if and only if $t \in [0,\omega)$. Let $t \in [0,\omega)$. Since $\pi_n \to \pi$ it follows that $x_n \pi_n t$ is defined for all n large enough and $x_n \pi_n t \to x\pi t$ as $n \to \infty$. Thus if $t < t_n$ for all n large enough, then $x\pi[0,t] \subset N$.

1.1 $t_n \to \infty$:
Suppose $\omega<\infty$. Then for some $t'<\omega$, $x\pi t' \notin N$. Since N is closed, $x_n\pi_n t' \notin N$ for n large. However, $t'<t_n$ for all such n, and this is a contradiction. It follows that $\omega=\infty$ and $x\pi[0,\infty)\subset N$, implying 1.1.

1.2 $t_n \to t_0 < \infty$.
Arguing as above we obtain that for every $\varepsilon>0$, $t_0-\varepsilon<\omega$ and $x\pi[0,t_0-\varepsilon]\subset N$. But then $t_0<\omega$ and N closed implies $x\pi[0,t_0]\subset N$.

2. Assume that N is $\{\pi_{n_m}\}$-admissible for every subsequence $\{\pi_{n_m}\}$ of $\{\pi_n\}$.

2.1 Let $t_n\to\infty$ as $n\to\infty$ and $x_n\pi_n[0,t_n]\subset N$ for all n. Let y_0 be a limit point of $\{x_n\pi_n t_n\}$, i.e. $x_n^0\pi_n^0 t_n^0 \to y_0$ as $n\to\infty$ for a subsequence $\{x_n^0\pi_n^0 t_n^0\}$ of $\{x_n\pi_n t_n\}$.
Since N is $\{\pi_n^0\}$-admissible, there is a subsequence $\{x_n^1\pi_n^1 t_n^1\}$ of $\{x_n^0\pi_n^0 t_n^0\}$ with $t_n^1\geq 1$ for $n\in\mathbb{N}$ and $x_n^1\pi_n^1(t_n^1-1)\to y_{-1}$ as $n\to\infty$.
Proceeding recursively we get for every $k\geq 1$ a subsequence $\{x_n^k\pi_n^k t_n^k\}$ of $\{x_n^{k-1}\pi_n^{k-1}t_n^{k-1}\}$ such that $t_n^k\geq k$ and $x_n^k\pi_n^k(t_n^k-k)\to y_{-k}$ as $n\to\infty$, where $y_{-k}\in X$.

We claim that $y_{-k}\pi t$ is defined for $t\in[0,k]$. In fact $y_{-k}\in N$ by our assumptions. If $y_{-k}\pi t$ is not defined on $[0,k]$, then there is a $0\leq t_0<k$ with $y_{-k}\pi t_0$ defined and $\notin N$. Hence $(x_n^k\pi_n^k(t_n^k-k))\pi_n^k t_0 \notin N$ for n large, a contradiction.

Define for $t\in[-k,0]$

$$\sigma_k(t) = y_{-k}\pi(t+k) \quad .$$

Then obviously σ_k is a solution of π in N with $\sigma_k(y_{-k})=y_0$. Moreover, if $k<k'$ then σ_k and $\sigma_{k'}$ coincide on $[-k,0]$. Then $\sigma(t):=\sigma_k(t), t\in[-k,0]$ defines a solution of π on $(-\infty,0]$ with $\sigma[\mathbb{R}^-]\subset N$ and $\sigma(0)=y_0$. Hence $y_0\in A_\pi^-(N)$ as claimed.

2.2 Suppose π_n does not explode in N for all n and $W\subset N$ with $A_\pi(N)\subset \mathrm{Int}W$. If our claim is not true, then w.l.o.g. we can assume that $A_{\pi_n}(N)\setminus(N\setminus\mathrm{Int}\,W)\neq\emptyset$ for all n. Let $K_n:=A_{\pi_n}(N)$, $K=A_\pi(N)$. Choose

$x_n\in K_n\cap(N\setminus\mathrm{Int}\,W)$. Then there are $y_n\in K_n$ $t_n\geq 0$ with $t_n\to\infty$ as $n\to\infty$ and $y_n\pi_n t_n=x_n$ for $n\in\mathbb{N}$. By admissibility we can assume that $y_n\pi_n t_n\to x_0$ for some x_0. Since $y_n\pi_n[0,t_n]\subset N$, 2.1 implies that $x_0\in A_\pi^-(N)$. As π_n does not explode in N, $x_n\pi_n t$ is defined for all $t\geq 0$. Moreover,

$x_n \pi_n t \in N$ for all $t \geq 0$. Hence $x_0 \pi t$ is defined and $\in N$ for all $t \geq 0$, i.e. $x_0 \in A_\pi^+(N)$.

Altogether we obtain $x_0 \in K \cap (N \setminus \mathrm{Int}\, W) = \emptyset$, a contradiction.

2.3 Assume $\pi_n \equiv \pi$ for all n. Let $\{x_n\}$ be a sequence in $A_\pi^-(N)$. Then there $y_n \in A_\pi^-(N)$, $t_n \geq 0$, $t_n \to \infty$ with $y_n \pi t_n = x_n$ for all n. By admissibility and part 2.1 of this proposition, $x_{n_m} = y_{n_m} \pi t_{n_m} \to x_0 \in A_\pi^-(N)$ as $m \to \infty$ for some subsequences $\{y_{n_m}\}$ and $\{t_{n_m}\}$. This proves that $A_\pi^-(N)$ is compact. If $\{x_n\}$ is a sequence in $K = A_\pi(N)$, then by what we have just proved. $x_{n_m} \to x_0 \in A_\pi^-(N)$ for a subsequence $\{x_{n_m}\}$.
Since π does not explode in N, $x_{n_m} \pi [0,\infty)$ is defined and $\subset N$. Hence $x_0 \pi [0,\infty)$ is defined and $\subset N$. That is, $x_0 \in A_\pi^+(N)$. This proves that K is compact.

1.5 Existence of isolating blocks

We are now in a position to state the basic

Theorem 5.1

Let $K \neq \emptyset$ be an isolated π-invariant set and let N be a strongly π-admissible isolating neighborhood of K. Then there exists an isolating block B, such that $K \subset B \subset N$.

Remark:

Theorem 5.1 remains true for $K = \emptyset$; just take $B = \emptyset$. Theorem 5.1 asserts the existence of isolating blocks for those invariant sets which admit strongly π-admissible isolating neighborhoods.
Before proceeding with the proof of Theorem 5.1, let us make a few simplifications:
First assume that π is a global semiflow, i.e., that $\omega_x = \infty$ for every $x \in X$. The reader will easily see that the proof of the general case is only notationally more complicated. Next assume that there is an open set U such that $K \subset U \subset \mathrm{Cl}\, U = N$. This can always be achieved by making N smaller, if necessary.
Define the following functions:

(i) $s^+ : N \to \mathbb{R} \cup \{\infty\}$, $s^+(x) = \sup\{t \mid x\pi[0,t] \subset N\}$,

(ii) $t^+ : U \to \mathbb{R} \cup \{\infty\}$, $t^+(x) = \sup\{t \mid x\pi[0,t] \subset U\}$,

(iii) $F : X \to [0,1]$, $F(x) = \min\{1, d(x, A^-(N))\}$,

(iv) $G : X \to [0,1]$, $G(x) = d(x,K)/(d(x,K) + d(x, X \setminus N))$,

(v) $g^+(x) = \inf\{(1+t)^{-1} G(x\pi t) \mid 0 \leq t < t^+(x)\}$,

(vi) $g^-(x) = \sup\{\alpha(t) F(x\pi t) \mid 0 \leq t \leq s^+(x)$, if $s^+(x) < \infty$, and $0 \leq t < \infty$ if $s^+(x) = \infty\}$

g^+ is defined on U, g^- is defined on N and $\alpha:[0,\infty)\to[1,2)$ is a monotone increasing C^∞-diffeomorphism.

Now the following proposition holds:

Proposition 5.2

Let the assumptions of Theorem 5.1 be satisfied and s^+, t^+, g^+ and g^- be as above. Then the following properties hold:

(1) *s^+ is upper-semicontinuous, t^+ is lower-semicontinuous.*
(2) *g^+ is upper-semicontinuous, and g^+ is continuous in a neighborhood of K.*
Moreover, if $g^+(x)\neq 0$, then

$$\dot{g}^+(x) = \liminf_{s\to 0^+}(1/s)(g^+(x\pi s)-g^+(x)) > 0.$$

On the other hand, if $g^+(x)=0$, then for every $t>0$, $x\pi t\in U$ and $g^+(x\pi t)=0$.
(3) *g^- is upper-semicontinuous. If $t^+=s^+$ on U, then g^- is continuous on U. Moreover, if $g^-(x)\neq 0$, then*

$$\dot{g}^-(x) = \limsup_{s\to 0^+}(1/s)(g^-(x\pi s)-g^-(x)) < 0.$$

On the other hand, if $g^-(x)=0$, then for every $0\leq t<s^+(x)$, $g^-(x)=0$.

Proof of the proposition:

(1) Let $x_0\in N$ and $\mu>s^+(x_0)$ be arbitrary. By the definition of s^+, there is a t_0, $s^+(x_0)<t_0<\mu$, such that $x_0\pi t_0\notin N$. By continuity of π, it follows that $x\pi t_0\notin N$, for all x in a small neighborhood of x_0. Hence $s^+(x)<t_0<\mu$ for all such x. This proves that s^+ is upper-semicontinuous.

Now let $x_0\in U$ and $\mu<t^+(x_0)$ be arbitrary. We can assume that $\mu>0$. Thus $x_0\pi[0,\mu]\subset U$. Therefore $x\pi[0,\mu]\subset U$ for all x close to x_0.
Hence $\mu<t^+(x)$ for all such x. It follows that t^+ is lower-semicontinuous.

(2) Let $x_0\in U$ and $\mu>g^+(x_0)$ be arbitrary. By the definition of g^+, $(1+t_0)^{-1}G(x_0\pi t_0)<\mu$ for some $t_0<t^+(x_0)$. Since t^+ is lower-semicontinuous, $t_0<t^+(x)$ and $(1+t_0)^{-1}G(x\pi t_0)<\mu$ for all x close to x_0. This implies $g^+(x)<\mu$ for all such x, proving that g^+ is upper-semicontinuous. To prove that g^+ is continuous in a neighborhood of K, we need the following lemma:

Lemma 5.3

If g^+ is not lower-semicontinuous at a point $a_0 \in U$, then there exists a sequence $\{a_m\} \subset U$, $a_m \to a_0$ as $m \to \infty$, and a sequence $\{t_m\} \subset \mathbb{R}^+$ such that $t^+(a_0) < t_m < t^+(a_m)$ and $(1+t_m)^{-1} G(a_m \pi t_m) < g^+(a_0)$ for all m.

Proof of the lemma:

By our hypothesis, there exists a $\mu \in \mathbb{R}$ and a sequence $a_m \to a_0$ such that $g^+(a_0) > \mu > g^+(a_m)$ for all m. Therefore, there is a sequence t_m, $0 \le t_m < t^+(a_m)$, such that $(1+t_m)^{-1} G(a_m \pi t_m) < \mu < g^+(a_0)$ for all m. Taking subsequences if necessary, we can assume w.l.o.g. that $t_m \to t_0 \le \infty$.

Obviously $t^+(a_0) < \infty$, since otherwise $g^+(a_0) = 0$. Suppose that $t_0 \le t^+(a_0)$. Then $(1+t_m)^{-1} G(a_m \pi t_m) \to (1+t_0)^{-1} G(a_0 \pi t_0)$, which implies that $(1+t_0)^{-1} G(a_0 \pi t_0) \le \mu < g^+(a_0)$, a contradiction to the definition of g^+.

Hence, $t_0 > t^+(a_0)$, and therefore, $t_m > t^+(a_0)$ for m large enough.

The lemma is proved.

Now assume that g^+ is not lower-semicontinuous in any neighborhood of K. Then there is a sequence $\{x_n\} \subset U$, $x_n \to K$, such that g^+ is not lower-semicontinuous at x_n. Since K is compact by Theorem 4.5, we can assume that $x_n \to x_0 \in K$.

By Lemma 5.3, for every n there exist sequences $\{t_m^n\}$, $\{a_m^n\}$, $m \ge 1$, such that $a_m^n \underset{m\to\infty}{\to} x_n$, $t^+(x_n) < t_m^n < t^+(a_m^n)$ and $(1+t_m^n)^{-1} G(a_m^n \pi t_m^n) < g^+(x_n)$ for all $n, m \ge 1$. Since $t^+(x_n) < \infty$ for every n, $x_n \pi t^+(x_n) \in \partial U$, and therefore, there is an m_n such that $d(x_n, a_{m_n}^n) < 2^{-n}$ and $d(a_{m_n}^n \pi t^+(x_n), \partial U) < 2^{-n}$. Let $y_n = a_{m_n}^n$ and $t_n = t_{m_n}^n$ for all n. It follows that for all n:

(1) $d(x_n, y_n) < 2^{-n}$, $d(y_n \pi t^+(x_n), \partial U) < 2^{-n}$;

(2) $t^+(x_n) < t_n < t^+(y_n)$;

(3) $(1+t_n)^{-1} G(y_n \pi t_n) < g^+(x_n)$.

Set $s_n = \dfrac{t^+(x_n)}{2}$. Since $x_n \to x_0 \in K$, $t^+(x_0) = \infty$, therefore by (1) of this proposition, $t^+(x_n) \to \infty$. Hence $x_n \pi [0, s_n] \subset N$ and $s_n \to \infty$. By admissibility of N, the sequence $\{x_n \pi s_n\}$ is precompact.

Hence, w.l.o.g., $x_n \pi s_n \to y_0 \in A^-(N)$ (cf. Theorem 4.5). Theorem 4.5 also implies that since $(x_n \pi s_n) \pi [0, s_n] \subset N$ and $s_n \to \infty$, therefore $y_0 \in A^+(N)$. Hence $y_0 \in A^-(N) \cap A^+(N) = K$. From (3) we obtain easily

(4) $G(y_n \pi t_n) < (1+t_n)(1+s_n)^{-1} G(x_n \pi s_n)$ for all n.

Now, by (2), it follows that $y_n \pi [0, t^+(x_n)] \subset N$ and $t^+(x_n) \to \infty$. Again admissibility of N implies that w.l.o.g. $y_n \pi t^+(x_n) \to z_0 \in A^-(N)$. By (1), $z_0 \in A^-(N) \cap \partial U$.

Now there are two possible cases:

1. Case: $\{t_n - t^+(x_n)\}$ is bounded.
Then, w.l.o.g. $t_n - t^+(x_n) \to t_0 < \infty$.
Then

$$(1+t_n)(1+s_n)^{-1} = \frac{1+t^+(x_n) + (t_n - t^+(x_n))}{1+(1/2)t^+(x_n)}$$

is bounded. Now $G(x_n \pi s_n) \to G(y_0) = 0$, as $y_0 \in K$, thus by (4), $G(y_n \pi t_n) \to 0$ as $n \to \infty$. Moreover, $(y_n \pi t^+(x_n)) \pi [0, t_n - t^+(x_n)] \subset N$ for all n, and $y_n \pi t^+(x_n) \to z_0$. Theorem 4.5 implies that $z_0 \pi [0, t_0] \subset N$. Now $y_n \pi t_n \to z_0 \pi t_0$ and so $G(z_0 \pi t_0) = 0$, i.e., $z_0 \pi t_0 \in K$. It follows that $z_0 \pi [0, \infty) \subset N$, i.e., $z_0 \in A^+(N)$.
On the other hand, $z_0 \in A^-(N) \cap \partial U$, and therefore $z_0 \in K \cap \partial U$ a contradiction, since $K \cap \partial U = \emptyset$.

2. Case: $\{t_n - t^+(x_n)\}$ is unbounded.
Then, w.l.o.g., $t_n - t^+(x_n) \to \infty$.
Hence $(y_n \pi t^+(x_n)) \pi [0, t_n - t^+(x_n)] \subset N$ and $y_n \pi t^+(x_n) \to z_0$.

Theorem 4.5 implies $z_0 \in A^+(N)$ which, as before, yields a contradiction. This proves that, indeed, g^+ is continuous in a neighborhood of K.

Now suppose that $x \in U$ is such that $g^+(x) \neq 0$. Then $t^+(x) < \infty$. Let $s > 0$ be small. Then $t^+(x \pi s) = t^+(x) - s$, and there is a $t_1 = t_1(s)$, $s \le t_1 \le t^+(x)$, such that $g^+(x \pi s) = (1+t_1 - s)^{-1} G(x \pi t_1)$. This implies

$$(1/s)(g^+(x \pi s) - g^+(x)) =$$

$$= (1/s)[(1+t_1-s)^{-1} G(x \pi t_1) - g^+(x)] \geq$$

$$\geq (1/s)[(1+t_1-s)^{-1} - (1+t_1)^{-1}] G(s \pi t_1) \geq$$

$$\geq (1+t_1)^{-2} G(x \pi t_1) \geq (1+t^+(x))^{-1} g^+(x) > 0 .$$

Hence $\dot{g}^+(x) > 0$. The last statement of (2) is obvious.

(3) Suppose g^- is not upper-semicontinuous at a point $x_0 \in N$. Then there are $\mu > 0$, $\{x_n\} \subset N$, such that $x_n \to x_0$ as $n \to \infty$ and $g^-(x_0) < \mu < g^-(x_n)$. Consequently there are $0 \le t_n < \infty$, $t_n \le s^+(x\ )$, such that

$g^-(x_0)<\mu<\alpha(t_n)\cdot F(x_n\pi t_n)$ for all $n\geq 1$.

We may assume $t_n\to t_0\leq\infty$.

Two cases are possible:

1. Case: $t_0=\infty$.

Then Theorem 4.5 implies that, w.l.o.g. $x_n\pi t_n\to z_0\in A^-(N)$. Hence $\alpha(t_n)\cdot F(x_n\pi t_n)\to 2\cdot F(z_0)=0$, which implies $g^-(x_0)<\mu\leq 0$, a contradiction.

2. Case: $t_0<\infty$.

Then $x_0\pi[0,t_0]\subset N$ and $\alpha(t_n)\cdot F(x_n\pi t_n)\to\alpha(t_0)\cdot F(x_0\pi t_0)$. Hence $t_0\leq s^+(x_0)$ and $g^-(x_0)<\mu\leq\alpha(t_0)\cdot F(x_0\pi t_0)\leq g^-(x_0)$, a contradiction.
We have proved that g^- is upper-semicontinuous.

Now suppose that $t^+=s^+$ on U and $x_0\in U$ be arbitrary. If $g^-(x_0)>\mu$, then for some $0\leq t_0<\infty$, $t_0<s^+(x_0)$, $\alpha(t_0)\cdot F(x_0\pi t_0)>\mu$.
By our assumption and part (1) this proposition, s^+ is continuous on U. Hence for all x close to x_0, we get $t_0<s^+(x)$ and $\alpha(t_0)\cdot F(x\pi t_0)>\mu$.

This implies $g^-(x)>\mu$ for all such x, and proves that g is lower semicontinuous on U.
Now let $x\in N$ be such that $g^-(x)\neq 0$.
There are two possible cases:

<u>1. Case</u>: $s^+(x)=\infty$.

Theorem 4.5 implies that $x\pi t\to K\subset A^-(N)$ as $t\to\infty$, and therefore $F(x\pi t)\to 0$ as $t\to\infty$. Moreover, since $g^-(x)\neq 0$, it follows easily that $x\notin A^-(N)$, and hence $F(x)\neq 0$. It follows that for some small $h>0$, $\inf_{0\leq s\leq h} F(x\pi s)=\mu>0$, and for some large $a>0$, $F(x\pi s)<\mu/4$ for $s\geq a$. Hence for every s with $0\leq s\leq h$, there is a $t_1=t_1(s)$ such that $s\leq t_1\leq a$ and such that $g^-(x\pi s)=\alpha(t_1-s)F(x\pi t_1)$. It easily follows that $F(x\pi t_1)\geq(1/2)\cdot F(x\pi s)\geq\mu/2$.

Then

$$\begin{aligned}&(1/s)(g^-(x\pi s)-g^-(x))\leq\\ &\leq(1/s)(\alpha(t_1-s)-\alpha(t_1))\cdot F(x\pi t_1)\leq\\ &\leq -(1/s)\cdot(\mu/2)\cdot\int_{t_1-s}^{t_1}\dot\alpha(u)\,du\leq-(\nu\cdot\mu/2)\quad,\end{aligned}$$

where $\nu=\inf_{0\leq u\leq a}\dot\alpha(u)>0$.

It follows that $\dot g^-(x)<0$.

2. Case: $s^+(x)<\infty$.

This case is proved by replacing a by $s^+(x)$ in the 1. case. The last statement of (3) is obvious.
The proposition is proved.
We now obtain the following

Lemma 5.4

Under the assumption of Theorem 5.1 set $H_{\varepsilon_1,\varepsilon_2} = \{x \in U | g^+(x)<\varepsilon_1,\ g^-(x)<\varepsilon_2\}$, where $\varepsilon_1,\varepsilon_2>0$ are arbitrary. Then $H_{\varepsilon_1,\varepsilon_2}$ is open, and if ε_1 and ε_2 are small enough, then $\mathrm{Cl}H_{\varepsilon_1,\varepsilon_2} \subset U$, g^+ is continuous on $\mathrm{Cl}H_{\varepsilon_1,\varepsilon_2}$, and $\mathrm{Cl}H_{\varepsilon_1,\varepsilon_2}$ is an isolating neighborhood of K.

Proof of Lemma 5.4:

Since both g^+ and g^- are upper-semicontinuous, $H_{\varepsilon_1,\varepsilon_2}$ is open for every $\varepsilon_1,\varepsilon_2>0$.
Also, if $x \in K$, then $g^+(x)=g^-(x)=0$, so that $K \subset H_{\varepsilon_1,\varepsilon_2}$. Suppose now that $\{x_n\}$ is a sequence such that $g^+(x_n)\to 0$ and $g^-(x_n)\to 0$ as $n\to\infty$.
We claim that there is a subsequence of $\{x_n\}$ converging to some $x \in K$.
In fact, $g^-(x_n)\to 0$ implies that $F(x_n)\to 0$ as $n\to\infty$.
By compactness of $A^-(N)$, there is a subsequence of $\{x_n\}$, denoted by $\{x_n\}$ again, and $x \in A^-(N)$ such that $x_n \to x$ as $n\to\infty$.
Two cases are possible:

1. Case: $\{t^+(x_n)\}$ is bounded by some $M<\infty$.
Then $g^+(x_n) \geq (1+M)^{-1} \inf\{G(x_n \pi t) | 0 \leq t \leq t^+(x_n)\} = (1+M)^{-1} G(x_n \pi t_n)$, for some $0 \leq t_n \leq t^+(x_n)$.
We may assume that $t_n \to t_0 < \infty$.
Theorem 4.5 implies that $x\pi[0,t_0] \subset N$ and $G(x_n \pi t_n) \to G(x \pi t_0)$. Since $g^+(x_n)\to 0$, this implies that $G(x\pi t_0)=0$ and hence $x\pi t_0 \in K$. It follows that $x \in A^+(N)$. But $x \in A^-(N)$; hence $x \in K$.

2. Case: $\{t^+(x_n)\}$ is unbounded.
Then we may assume that $t^+(x_n)\to\infty$, and therefore, by Theorem 4.5 $x \in A^+(N)$.
Therefore again, $x \in A^+(N) \cap A^-(N) = K$.
Our claim is proved.

Suppose now that $ClH_{\varepsilon_1,\varepsilon_2}\not\subset U$ for all $\varepsilon_1,\varepsilon_2>0$. Then there are sequences $\varepsilon_n\to 0$, $y_n\in ClH_{\varepsilon_n,\varepsilon_n}\cap\partial U$.

There are $x_n\in H_{\varepsilon_n,\varepsilon_n}$, $d(x_n,y_n)<\varepsilon_n$.

It follows that $g^+(x_n)$ and $g^-(x_n)\to 0$ as $n\to\infty$. Our claim implies that, w.l.o.g., $x_n\to x\in K$. But then $y_n\to x$ as $n\to\infty$. We thus obtain that $x\in K\cap\partial U$, a contradiction.

Hence, indeed, $ClH_{\varepsilon_1,\varepsilon_2}\subset U$ for all $\varepsilon_1,\varepsilon_2>0$ small enough.

If there is no $\varepsilon>0$, such that g^+ is continuous on $ClH_{\varepsilon,\varepsilon}$ then there are $\varepsilon_n\to 0$, $y_n\in ClH_{\varepsilon_n,\varepsilon_n}$, such that g^+ is not continuous at y_n. Arguing as above we may assume w.l.o.g. that $y_n\to x\in K$. But g^+ is continuous on a neighborhood of K, a contradiction. Hence g^+ is continuous on $ClH_{\varepsilon_1,\varepsilon_2}$, for all $\varepsilon_1,\varepsilon_2>0$ small enough.

The last statement of the lemma is obvious.

We can now prove Theorem 5.1:

Choose $\varepsilon>0$ so small that $ClH_{2\varepsilon,2\varepsilon}\subset U$ and g^+ is continuous on $ClH_{2\varepsilon,2\varepsilon}$, set $\tilde{U}=H_{\varepsilon,\varepsilon}$, $\tilde{N}=ClH_{\varepsilon,\varepsilon}$.

Then $K\subset\tilde{U}\subset Cl\tilde{U}=\tilde{N}\subset U$, and $\tilde{N}$ strongly π-admissible isolating neighborhood of K. Now define the functions $\tilde{s}^+,\tilde{t}^+,\tilde{g}^+,\tilde{g}^-$ as in (i)-(vi), where $\tilde{U}$ and $\tilde{N}$ is replaced by $\tilde{U}$ and $\tilde{N}$, respectively. For example $\tilde{s}^+(x)=\sup\{t\,|\,x\pi[0,t]\subset\tilde{N}\}$.

Of course, Proposition 5.2 and Lemma 5.4 hold for $\tilde{s}^+,\tilde{t}^+,\tilde{g}^+$ and $\tilde{g}^-$.

We will show that $\tilde{t}^+=\tilde{s}^+$ on $\tilde{U}$.

Indeed, $\tilde{t}^+\leq\tilde{s}^+$ on $\tilde{U}$, of course. Suppose $\tilde{t}^+(x)<\tilde{s}^+(x)$ for some $x\in\tilde{U}$. It follows that $\tilde{t}^+(x)<\infty$ and $y=x\pi\tilde{t}^+(x)\in\partial\tilde{U}$. By Lemma 5.4 $\partial\tilde{U}\subset U$, and hence $y\in U$.

Consequently $g^+(y)$ and $g^-(y)$ are defined and either $g^+(x)\geq\varepsilon$ or else $g^-(y)\geq\varepsilon$. If $g^-(y)\geq\varepsilon$, then, by Proposition 5.2, $g^-(x)\neq 0$ and $g^-(x)>g^-(y)\geq\varepsilon$. Hence $g^-(x)>\varepsilon$ which contradicts $x\in\tilde{U}$.

It follows that $g^+(y)\geq\varepsilon$. Now, since $y\in\partial\tilde{U}$, there are $y_n\in\tilde{U}$, $y_n\to y$. Since g^+ is continuous on $ClH_{2\varepsilon,2\varepsilon}$ and $Cl\tilde{U}\subset ClH_{2\varepsilon,2\varepsilon}$, it follows that $g^+(y_n)\to g^+(y)$. Hence $g^+(y)=\varepsilon$, i.e., $y\in H_{2\varepsilon,2\varepsilon}$. Choose t such that $\tilde{t}^+(x)<t<\tilde{s}^+(x)$ and such that $x\pi[0,t]\subset H_{2\varepsilon,2\varepsilon}$. This is possible since $H_{2\varepsilon,2\varepsilon}$ is open. It follows that $g^+(x\pi t)>g^+(y)=\varepsilon$.

By the continuity of g^+ on $ClH_{2\varepsilon,2\varepsilon}$, it follows that there is an open set V, $x\pi t\in V\subset H_{2\varepsilon,2\varepsilon}$, such that $g^+(z)>\varepsilon$ for $z\in V$. Hence $V\cap H_{\varepsilon,\varepsilon}=\emptyset$, and so $x\pi t\notin ClH_{\varepsilon,\varepsilon}=\tilde{N}$. But this implies $t>\tilde{s}^+(x)$, a contradiction.

This proves that indeed, $\tilde{t}^+=\tilde{s}^+$ on $\tilde{U}$. Proposition 5.2 now implies that $\tilde{g}^-$ is continuous on $\tilde{U}$. Now apply Lemma 5.4 to $\tilde{U},\tilde{N}$.
Then for $\delta>0$ sufficiently small the set $\tilde{H}_\delta=\{x\in\tilde{U} \mid \tilde{g}^+(x)<\delta,\ \tilde{g}^-(x)<\delta\}$ has the following properties:

(1) $\tilde{H}_\delta$ is open, $K\subset\tilde{H}_\delta$, $\mathrm{Cl}\tilde{H}_\delta\subset U$;
(2) $\tilde{g}^+$ and $\tilde{g}^-$ are continuous on $\mathrm{Cl}\tilde{H}_\delta$.

Choose such $\delta>0$, and let $0<\delta_1,\ \delta_2<\delta/2$ be arbitrary. Define

$$B = B_{\delta_1,\delta_2}=\mathrm{Cl}\{x\in\tilde{U} \mid \tilde{g}^+(x)<\delta_1\ ,\ \tilde{g}^-(x)<\delta_2\} \tag{1}$$

We will show that $B\subset\tilde{H}_\delta$. In fact, if $x\in B$, then $x\in\mathrm{Cl}\tilde{H}_\delta$, and, moreover, there is a sequence $x_n\to x$, as $n\to\infty$, and such that $\tilde{g}^+(x_n)<\delta_1$ and $\tilde{g}^-(x_n)<\delta_2$ for all n. Since $\tilde{g}^+$ and $\tilde{g}^-$ are continuous on $\mathrm{Cl}\tilde{H}_\delta$, it follows that $\tilde{g}^+(x)\le\delta/2<\delta$ and $\tilde{g}^-(x)\le\delta/2<\delta$, i.e., $x\in\tilde{H}_\delta$.

Now set

$$b^- = \{x\in\partial B \mid \tilde{g}^+(x)=\delta_1,\ \tilde{g}^-(x)<\delta_2\} \tag{2}$$

$$b^+ = \{x\in\partial B \mid \tilde{g}^+(x)<\delta_1,\ \tilde{g}^-(x)=\delta_2\} \tag{3}$$

$$\tau = \{x\in\partial B \mid \tilde{g}^+(x)=\delta_1,\ \tilde{g}^-(x)=\delta_2\}. \tag{4}$$

It follows that $\partial B=b^-\cup b^+\cup\tau$.

Using the continuity of $\tilde{g}^+$ and $\tilde{g}^-$ on H_δ as well as the fact that $\tilde{g}^+$ (resp. $\tilde{g}^-$) increases (resp. decreases) along solutions of π (by Proposition 5.2) it is easily proved that:

every point of b^- is a strict egress point
every point of b^+ is a strict ingress point,

and every point of τ is a bounce-off point.

Hence $b^-\subset B^e$, $b^+\subset B^i$, $\tau\subset B^b$, which in fact, implies that $B^-=b^-\cup\tau$ and therefore, B^- is closed. This proves that B is an isolating block.
The theorem is proved.

Corollary 5.5

Suppose that all assumptions of Theorem 5.1 are satisfied, and, in addition, that

$A^-(N) = A(N) = K$.

Then there exists an isolating block B , $K\subset B\subset N$ with $B^- = \emptyset$.

Remark: The corollary is also valid for $K=\emptyset$, taking $B=\emptyset$.

Proof: Choose $\delta>0$ such that for $0<\delta_1,\delta_2<\delta/2$. $B=B_{\delta_1,\delta_2}$ defined in (1) of the proof of Theorem 5.1 is an isolating block. Pick any $0<\delta_1<\delta/2$. We will show that there is a $0<\delta_2<\delta/2$ such that whenever $x\in\tilde{U}$ and $\tilde{g}^+(x)=\delta_1$, then $\tilde{g}^-(x)>\delta_2$. With this choice of δ_1,δ_2, b^- and τ in (2) and (4) above are empty, so $B^-_{\delta_1,\delta_2}=\emptyset$, as claimed.
If our claim is not true, then there is a sequence $x_n\in\tilde{U}$ with $\tilde{g}^+(x_n)=\delta_1$ and $\tilde{g}^-(x_n)\to 0$ as $n\to\infty$.
This implies $\mathrm{dist}(x_n, A^-(\tilde{N}))\to 0$ as $n\to\infty$.

Since $K\subset\tilde{N}\subset N$ and $A^-(N)=K$, it follows that $A^-(\tilde{N})=K$. Since K is compact, we may assume that $x_n\to x\in K$ as $n\to\infty$. By the definition of $\tilde{g}^+$, it follows that $\tilde{g}^+(x_n)\to\tilde{g}^+(x)=0$, a contradiction.
The corollary is proved.

1.6 Homotopies and inclusion induced maps

In this section we will recall a few concepts from the homotopy theory which we need for the construction of the Morse index.

If Y is a topological space and $A\subset Y$ is a subspace, then (Y,A) is called a topological pair. In the special case where $A=\emptyset$, we identify (Y,A) with Y. Moreover, if $A=\{y_0\}$, $y_0\in Y$, then we identify (Y,A) with (Y,y_0) and call (Y,y_0) a pointed-space with base-point y_0.

If (Y,A), (Z,B) are two topological pairs, then a morphism $f:(Y,A)\to(Z,B)$ is a continuous map (denoted by the same symbol f) $f:Y\to Z$ such that $f(A)\subset B$.

The class of all topological pairs (resp. all pointed spaces) together with all the corresponding morphisms defines, in an obvious way, a category, denoted by TP (resp. by T*).

If $f:(Y,A)\to(Z,B)$ and $g:(Y,A)\to(Z,B)$ are two morphisms, then f is called homotopic to g (we write $f\sim g$), if there is a continuous mapping $H:Y\times[0,1]\to Z$ such that $H(A\times[0,1])\subset B$, and such that $H(x,0)\equiv f(x)$ and $H(x,1)\equiv g(x)$ for all $x\in Y$. H is called a homotopy from f to g.

Obviously, $\sim$ is an equivalence relation on the set of all morphisms from (Y,A) to (Z,B).
Moreover, if (W,C) is a third pair and $\tilde{f}$ and $\tilde{g}$ are two morphisms from (Z,B) to (W,C) and both $f\sim g$ and $\tilde{f}\sim\tilde{g}$, then $\tilde{f}\circ f\sim\tilde{g}\circ g$.
If (Y,A) and (Z,B) are two topological pairs, then (Y,A) is called homotopy equivalent to (Z,B) (we write $(Y,A)\sim(Z,B)$) if there are two

morphisms $f:(Y,A)\to(Z,B)$ and $g:(Z,B)\to(Y,A)$ such that $f\circ g\sim 1_{(Z,B)}$ and $g\circ f\sim 1_{(Y,A)}$, where $1_{(Y,A)}$ (resp. $1_{(Z,B)}$) are the identity morphisms on (Y,A) (resp. on (Z,B)).

$\sim$ is easily seen to be an equivalence relation on the class of all topological pairs (resp. on the class of all pointed spaces).

If (Y,A) is a topological pair (resp. if (Y,y_0) is a pointed space) then we denote by $[(Y,A)]$ (resp. by $[(Y,y_0)]$) the corresponding equivalence class, i.e., the class of all topological pairs (resp. the class of all pointed spaces) which are homotopy equivalent to (Y,A) (resp. to (Y,y_0)). We call $[(Y,A)]$ (resp. $[(Y,y_0)]$) the _homotopy type of_ (Y,A) (resp. (Y,y_0)).

Moreover, if $f:(Y,A)\to(Z,B)$ (resp. $f:(Y,y_0)\to(Z,z_0)$) is a morphism of topological pairs (resp. of pointed spaces), then $[f]$ denotes the equivalence class of f, i.e., the set of all morphisms which are homotopy equivalent to f.

We will denote by HT^* the homotopy category of T^*, i.e. objects of HT^* are pointed spaces, whereas the morphisms of HT^* are equivalence classes $[f]$ of morphisms f of T^*.

Now let Z be a topological space and A,Y be subspaces of Z. Suppose first that $Y\cap A\neq\emptyset$. Define an equivalence relation on Y in the following way: $x\sim y$ if and only if $x=y$ or $x,y\in Y\cap A$. $\sim$ collapses $Y\cap A$ to one point. For $x\in Y$, let $[x]$ be the equivalence class of x. Hence $[x]=\{x\}$ if $x\notin Y\cap A$, and $[x]=Y\cap A$ for $x\in Y\cap A$. By Y/A we denote the set of all equivalence classes of $\sim$.

Let $q:Y\to Y/A$ be the quotient map, i.e., $q(x)=[x]$ for $x\in Y$.

We endow Y/A with the terminal topology with respect to q, i.e., let $U\subset Y/A$ be open if and only if $q^{-1}(U)$ is open in Y.

We treat Y/A as a pointed-space, choosing the equivalence class of $Y\cap A$ (denoted by $[A]$) to be the base-point.

If $Y\cap A=\emptyset$, then choose any point $p\notin Y$, give $Y\dot{\cup}\{p\}$ the sum topology, and let $Y/A=(Y\dot{\cup}\{p\})/\{p\}$. Obviously, in this case, Y/A is canonically homeomorphic to $Y\dot{\cup}\{p\}$.

We choose p to be the base-point of $Y/\emptyset$, but we write, as before, $p=[A]$. In this case, the map $q:Y\to Y/A$, $q(x)=[x]$ is such that the topology of Y/A is terminal with respect to q, i.e., $U\subset Y/A$ is open if and only if $q^{-1}(U)$ is open in Y. This is easily verified.
Note that, if $V\subset Y$, and $V\supset Y\cap A$ or $V\cap A=\emptyset$, then $q^{-1}(q(V))=V$. Hence, in

this case, V is open in Y if and only if q(V) is open in Y/A.

This remark will be used in the proofs later on.

Let us now introduce inclusion induced maps:

Definition 6.1

Let Y be topological space and A, B ,C,D be subsets of Y. Let $q_A:A\to A/B$ and $q_C:C\to C/D$ be quotient maps.

A map $j:A/B\to C/D$ is called inclusion induced, if the following properties hold:

(1) $A\cap B$ is closed in A, $C\cap D$ is closed in C.

(2) $A\setminus B\subset C$ and $A\cap B\cap C\subset D$.

(3) If $z\in A/B$ then

$$j(z) = \begin{cases} q_C(x) \text{ if } z=q_A(x) \;, & x\in A\setminus B \\ [D] \quad \text{otherwise} . \end{cases}$$

Remark:

Note that an inclusion induced map is base-point preserving. An inclusion induced map is called admissible if j is continuous and $Cl(A\setminus B)\cap B\cap C\subset D$.

The following proposition holds:

Proposition 6.2.

If $j:A/B\to C/D$ is inclusion induced, and if $Cl(A\setminus B)\cap A\subset C$, then j is continuous.

Proof:

Define $G:A\to C/D$ as

$$G(x) = \begin{cases} q_C(x) & \text{if } x\in A\setminus B \\ [D] & \text{else} \end{cases}$$

If $x\in A\cap B$, then $G(x)=[D]$. Hence j is the unique base-point preserving map satisfying $G=j\circ q_A$. Moreover, j is continuous if and only if G is. Hence we only need to show that G is continuous.

Let $x_0\in A$ and U be open in C/D such that $G(x_0)\in U$. There is an open set W in Y such that $W\cap C=q_C^{-1}(U)$.

1. Case: $x_0\in A\setminus B$.

Define $\Omega=W\cap(A\setminus B)$. Since $A\cap B$ is closed in A, Ω is open in A. Moreover, since $G(x_0)=q_C(x_0) \in U$, $x_0 \in W\cap C$, which implies $x_0 \in \Omega$. Let $x \in \Omega$. Then $x \in W\cap(A\setminus B)\subset W\cap C$, and $G(x)=q_C(x) \in q_C(W\cap C)\subset U$. This proves continuity of G at x_0 in this case.

2. Case: $x_0 \in A\cap B$. Then $C\cap D\subset W$.

Define $\Omega=(W\cup(X\setminus Cl(A\setminus B))\cap A$. Ω is open in A. Let $x \in A\cap B$. Then, either $x \notin Cl(A\setminus B)$ and hence $x \in \Omega$ or else assume $x \in Cl(A\setminus B)$. Since $Cl(A\setminus B)\cap A\subset C$, it follows that $x \in C$ hence (as $A\cap B\cap C\subset D$) we obtain $x\in C\cap D\subset W$. Therefore, $A\cap B\subset\Omega$. It follows that $x_0 \in \Omega$.

Let $x \in \Omega$. Assume first that $x \in A\setminus B$.
Then $x \in C$, by Definition 6.1. Moreover, $G(x)=q_C(x)$. Hence $x \in Cl(A\setminus B)\cap A\subset C$, i.e., $x \in W\cap C$, and $G(x) \in q_C(W\cap C)=U$. Now assume $x \in A\cap B$.
Then $G(x)=[D] \in U$.
The proposition is proved.

1.7 Index and quasi-index pairs.

Given an isolating block B, the pair $\langle B,B^-\rangle$ enjoys certain properties which are crucial for the definitions of the Morse and the homotopy indices. These properties are generalized and abstracted in the definitions of index and quasi-index pairs. Intuitively, index and quasi-index pairs are obtained by squeezing or stretching the pair $\langle B,B^-\rangle$ (see Fig. 3).

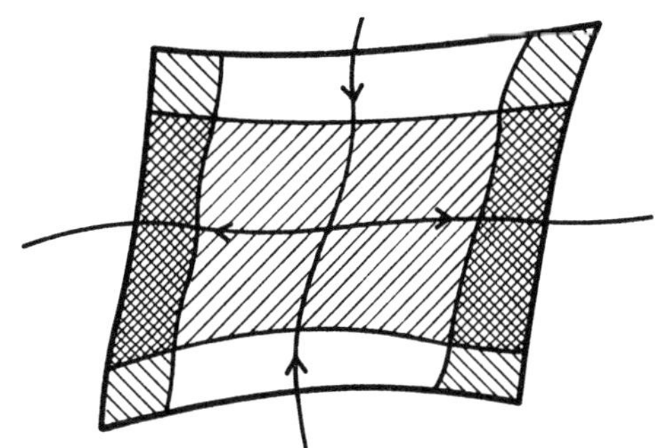

N_1 is the area shaded /////
N_2 is the area shaded \\\\\

Figure 3

In this section, as usual, X is a metric space and π is a local semiflow on X.
Let N,Y be subsets of X, $Y\subset N$. Y is called N-positively invariant (rel. to π), if whenever $x \in X$, $t\geq 0$ are such that $x\pi[0,t]\subset N$ and $x \in Y$,

then $x\pi[0,t]\subset Y$. In other words, a solution starting in Y cannot leave N without leaving Y.

Now let N_1, N_2, N be subsets of X, $N_2 \subset N$.
For $t \geq 0$, define

$$N_1^t = \{x \in X \mid \text{there is a } y \in X,\ 0 \leq t < \omega_y, \text{ and such that}$$
$$y\pi[0,t] \subset N_1 \text{ and } y\pi t = x\}.$$

In other words, N_1^t is the image under π of those points y the solution through which remains in N_1 for time t.

$$N_2^{-t} = N_2^{-t}(N) = \{x \in X \mid \text{there is a } t',\ 0 \leq t' \leq t,$$
$$\text{such that } t' < \omega_x,\ x\pi[0,t'] \subset N,$$
$$\text{and } x\pi t' \in N_2\}.$$

The solution through points in $N_2^{-t}(N)$ hits N_2 at a time $t' \leq t$ and does not leave N before.

We now arrive at the following definition:

<u>Definition 7.1</u>

Let K be an isolated invariant set and N be an isolating neighborhood of K.
An ordered pair $\langle N_1, N_2\rangle$ of sets is called an <u>index pair</u> in N if:

(1) N_1 and N_2 are closed subsets of N, and N_1, N_2 are N-positively invariant.
(2) $K \subset \mathrm{Int}(N_1 \setminus N_2)$.
(3) If $x \in N_1$ and $x\pi[0,\omega_x) \not\subset N$, then $x \in N_2^{-t}(N)$ for some $t \geq 0$.

<u>Remark</u>:

(3) means that if $x \in N_1$ and $x\pi t' \notin N$ for some $t' > 0$, then there is a $t < t'$ such that $x\pi[0,t] \subset N$ and $x\pi t \in N_2$.
If B is an isolating block for K, then the pair $\langle B, B^-\rangle$ is easily seen to be an index pair in N=B.
Moreover, for every $t \geq 0$, $\langle B, N_2\rangle$ is an index pair in N=B, where $N_2 = (B^-)^{-t}(N)$. This is trivially verified using Definition 7.1.
If π is generated by an ODE on a finite-dimensional space, then for every index pair $\langle N_1, N_2\rangle$ in N and every $t, s \geq 0$, $\langle N_1^s, N_2^{-t}(N)\rangle$ is also easily seen to be an index pair in N. This is due to the fact that $T(t)x = x\pi t$ is a local homeomorphism in this case.

For general semiflows, this latter property is no longer true and this justifies the introduction of the following, technical concept:

Definition 7.2

Let K and N be as in Definition 7.1.

An ordered pair $\langle N_1, N_2\rangle$ of subsets of N is called a quasi-index pair in N if there is a set $\tilde{N}_1$ such that

(1) $\langle \tilde{N}_1, N_2\rangle$ is an index pair in N, $\tilde{N}_1 \supset N_1 \setminus N_2$, and for some $\tilde{s} \geq 0$, $\tilde{N}_1^{\tilde{s}} \subset N_1$.
(2) Either N_1 is N-positively invariant, or else there exists a closed set M_1 which is $N \setminus N_2$-positively invariant, and there is a $\tilde{t} \geq 0$, such that $\tilde{N}_1 \supset M_1 \setminus N_2$ and $N_1 = M_1^{\tilde{t}}$.

The following proposition holds:

Proposition 7.3

If $\langle N_1, N_2\rangle$ is a quasi-index pair in N, then for every $t, s \geq 0$, $\langle N_1^s, N_2\rangle$ and $\langle (Cl(N_1^s))^t, N_2\rangle$ are both quasi-index pairs in N. Moreover, N_1 is $N \setminus N_2$-positively invariant and $X \setminus N_2$-positively invariant.

Proof:

Let $\tilde{N}_1$ and $\tilde{s} \geq 0$ be as in (1) of Definition 7.2. Obviously, $\tilde{N}_1 \supset N_1^s \setminus N_2$, since $\tilde{N}_1 \supset N_1 \setminus N_2$ and $N_1^s \subset N_1$. Let $x \in Cl(N_1^s) \setminus N_2$. Then there is a sequence $x_n \to x$, $x_n \in N_1^s$. Since N_2 is closed, we may assume that $x_n \in N_1^s \setminus N_2$ for all n. Hence $x_n \in \tilde{N}_1$ for all n, and therefore $x \in \tilde{N}_1$, as $\tilde{N}_1$ is closed. We obtain $(Cl(N_1^s))^t \setminus N_2 \subset \tilde{N}_1$. Moreover, let $r = s + t + \tilde{s}$.

Since $(Y^{s'})^{t'} = Y^{s'+t'}$ for any $t', s' \geq 0$ and any $Y \subset X$, it follows that

$$\tilde{N}_1^r = (\tilde{N}_1^{\tilde{s}})^{s+t} \subset N_1^{s+t} \subset N_1^s \cap (Cl(N_1^s))^t .$$

Hence (1) of Definition 7.2 is satisfied.

Now suppose first that N_1 is N-positively invariant. Then obviously, N_1^s is also N-positively invariant, and so (2) of Definition 7.2 is satisfied for $\langle N_1^s, N_2\rangle$ in this case.
Let $\tilde{M}_1 = Cl(N_1^s)$. Suppose $x \in \tilde{M}_1$ and $h \geq 0$ is such that $x\pi[0,h] \subset N \setminus N_2$. Then there are $x_n \in N_1^s$, $x_n \to x$ as $n \to \infty$. Since N_2 is closed, we may assume that $x_n\pi[0,h] \subset X \setminus N_2$ for all n. Since $N_1^s \setminus N_2 \subset \tilde{N}_1$ and $\langle \tilde{N}_1, N_2\rangle$ is an index pair in N, it follows that $x_n\pi[0,h] \subset N$ for all n. As N_1^s is N-positively invariant, it follows that $x_n\pi[0,h] \subset N_1^s$ for all n. This means that $x\pi[0,h] \subset Cl(N_1^s) = \tilde{M}_1$, i.e., $\tilde{M}_1$ is $N \setminus N_2$-positively invariant.
Finally $\tilde{N}_1 \supset \tilde{M}_1 \setminus N_2$ and $(Cl(N_1^s))^t = \tilde{M}_1^t$. Thus (2) of Definition 7.2 is satisfied for $\langle (Cl(N_1^s))^t, N_2\rangle$ in this case.

Now suppose that N_1 is not N-positively invariant. By (2) of Definition 7.2, there is a closed set M_1 and a $\tilde{t}\geq 0$, such that M_1 is $N\setminus N_2$-positively invariant, $M_1\setminus N_2\subset\tilde{N}_1$ and $M_1^{\tilde{t}}=N_1$. Hence $M_1^{\tilde{t}+s}=N_1^s$ and therefore (2) of Definition 7.2 is satisfied for $<N_1^s,N_2>$ in this case. Obviously $(Cl(N_1^s))^t=(Cl(M_1^{\tilde{t}+s}))^t$. Therefore (2) of Definition 7.2 is satisfied for $<(Cl(N_1^s))^t,N_2>$ if we can prove that $Cl(N_1^s)$ is $N\setminus N_2$-positively invariant.

In fact, suppose $x\in Cl(N_1^s)$ and $h\geq 0$ is such that $x\pi[0,h]\subset N\setminus N_2$. As above there is a sequence $x_n\in N_1^s$, $x_n\to x$ as $n\to\infty$ and such that $x_n\pi[0,h]\in X\setminus N_2$ for all n.
Therefore, $x_n\in M_1^{\tilde{t}+s}$, and since $M_1\setminus N_2\subset\tilde{N}_1\subset N$ and M_1 is $N\setminus N_2$-positively invariant, it follows that $x_n\pi[0,h]\subset M_1^{\tilde{t}+s}$ if we know that $x_n\pi[0,h]\subset N$. However, this latter statement follows from the fact that $x_n\in\tilde{N}_1$ and $<\tilde{N}_1,N_2>$ is an index pair in N.
Hence, indeed, $x_n\pi[0,h]\subset M_1^{\tilde{t}+s}=N_1^s$ for all n, and thus $x\pi[0,h]\subset Cl(N_1^s)$. This means that, in fact, $Cl(N_1^s)$ is $N\setminus N_2$-positively invariant.

All this proves that both $<N_1^s,N_2>$ and $<(Cl(N_1^s))^t,N_2>$ are quasi-index pairs in N.
The fact that N_1 is $N\setminus N_2$-positively invariant follows trivially, because either N_1 is N-positively invariant, or else $N_1=M_1^{\tilde{t}}$ where M_1 is $N\setminus N_2$-positively invariant.

If $x\in N_1$ and $h\geq 0$ is such that $x\pi[0,h]\subset X\setminus N_2$, then from $N_1\setminus N_2\subset\tilde{N}_1$ it follows that $x\in\tilde{N}_1$. Since $<\tilde{N}_1,N_2>$ is an index pair in N, we get $x\pi[0,h]\subset N\setminus N_2$. Since N_1 is $N\setminus N_2$-positively invariant, this means that $x\pi[0,h]\subset N_1$. Hence, indeed, N_1 is $X\setminus N_2$-positively invariant.

The proposition is proved.

Remarks:

1. In the sequel, if $<N_1,N_2>$ is a quasi-index pair in N and $s\geq 0$, then by N_2^{-s} we always mean $N_2^{-s}(N)$.
2. Let $<N_1,N_2>$ be a quasi-index pair in N and $<\tilde{N}_1,N_2>$ a corresponding index pair. If $x\in N_1$ and $x\pi[0,\omega_x)\not\subset N$, then $x\in N_2^{-t}(N)$ for some $t\geq 0$. In fact, if $x\in N_2$ then we are done. If $x\notin N_2$, then $x\in N_1\setminus N_2\subset\tilde{N}_1$, so the claim follows from (3) of Definition 7.1.
In other words, a solution starting in N_1 and leaving N_2 must first get to N_2. On the other hand, a solution starting in N_2 must leave N, if N is strongly π-admissible. In fact, in that case $x\in N_2$ and $x\pi[0,\omega_x]\subset N$ would imply $\omega_x=\infty$, $x\pi[0,\omega_x)\subset N$ and $x\pi t\to K$ as $t\to\infty$. Conse-

quently $K \cap N_2 \neq \emptyset$, which contradicts (2) of Definition 7.1. These arguments justify calling N_2 the exit ramp of the index or quasi-index pair $\langle N_1, N_2 \rangle$, a term suggested by James Yorke (oral communication).
3. Let $\langle N_1, N_2 \rangle$ be a quasi-index pair in N, and suppose that π does not explode in N. Then it is easily proved that for every $t \geq 0$, $N_2^{-t} = N_2^{-t}(N)$ is closed and, moreover, $N_2^{-t} \subset N_2^{-s}$ and $N_1^s \subset N_1^t$ for $0 \leq t < s$. This easily implies by Proposition 6.2 that there is a (uniquely determined) admissible inclusion induced map $e: N_1^s/N_2 \to N_1/N_2^{-t}$, where $s, t \geq 0$ are arbitrary. We will use this remark in later sections.

1.8 Some special maps used in the construction of the Morse index

The following technical results are needed for the definition of the Morse index:

Proposition 8.1

Let N be closed in X. Assume that π does not explode in N. Let $\langle N_1, N_2 \rangle$ be a quasi-index pair in N. Let $M_2 \subset N$ be closed, N-positively invariant, $N_2 \subset M_2$.
Let $s \geq 0$ and define $h: N_1/M_2 \times [0,1] \to N_1/M_2$ by

$$h(z,\sigma) = \begin{cases} q(x\pi\sigma s) & \text{if there is an } x \in N_1 \text{ with } x\pi[0,\sigma s] \cap N_2 = \emptyset \\ & \text{and } q(x) = z, \\ [M_2] & \text{otherwise}. \end{cases}$$

Here $q: N_1 \to N_1/M_2$ is the quotient map. (i.e., $q(x) = [x]$).
Then h is a well-defined, continuous and base-point preserving mapping.

Proof:

Write $I = [0,1]$.
To show that h is well-defined, let $(z,\sigma) \in N_1/M_2 \times I$ be given. Assume first $N_1 \cap M_2 \neq \emptyset$.
If $z \neq [M_2]$, then $q^{-1}(\{z\}) = \{x\}$ for some $x \in N_1 \setminus M_2$. If $x\pi[0,\sigma s] \cap N_2 = \emptyset$, then, by Proposition 7.3 $x\pi\sigma s \in N_1$, so $h(z,\sigma)$ is uniquely defined in this case. If $x\pi t \in N_2$ for some $t < \omega_x$, $t \in [0,\sigma s]$, then $h(z,\sigma) = [M_2]$, so h is uniquely defined, again. If $z = [M_2]$, then $q^{-1}(\{z\}) = N_1 \cap M_2$. Let $x \in N_1 \cap M_2$ be arbitrary. Suppose $x\pi[0,\sigma s] \cap N_2 = \emptyset$. Then, by Proposition 7.3 $x\pi[0,\sigma s] \subset N_1 \subset N$. Since M_2 is N-positively invariant, $x\pi\sigma s \in N_1 \cap M_2$. Thus $h(z,\sigma) = [M_2]$. If $x\pi t \in N_2$ for some $t < \omega_x, t \in [0,\sigma s]$, then, also, $h(z,\sigma) = [M_2]$.
Consequently h is well-defined and base-point preserving if $N_1 \cap M_2 \neq \emptyset$.

Now suppose that $N_1 \cap M_2 = \emptyset$. Let $x \in N_1$. Let $t_0 = \sup\{t < \omega_x \mid x\pi[0,t] \cap N_2 = \emptyset\}$. Since N_2 is closed, it follows that $t_0 > 0$. Suppose that $t_0 < \infty$. Let $<\tilde{N}_1, \tilde{N}_2>$ be an index pair in N, corresponding to $<N_1, N_2>$. Then $x \in N_1 \setminus N_2 \subset \tilde{N}_1$ and so $x\pi[0,t_0) \subset N \setminus N_2$. Since π does not explode in N and N is closed, $x\pi t_0$ is defined. Moreover, $x\pi t_0 \in N_2$.

If N_1 is N-positively invariant, then it follows that $x\pi t_0 \in N_1$, i.e., $x\pi t_0 \in N_1 \cap N_2 \subset N_1 \cap M_2 = \emptyset$, a contradiction. If $N_1 = \tilde{M}_1^t$, M_1 as in Definition 7.2, then $x\pi[0,t_0) \subset M_1$, since N_1 is $N \setminus N_2$-positively invariant. Since M_1 is closed, $x\pi[0,t_0] \subset M_1$, but since $x \in N_1 = \tilde{M}_1^t$, we get $x\pi t_0 \in \tilde{M}_1^t = N_1$, a contradiction again.
It follows that $t_0 = \infty$. In other words, for every $x \in N_1$ and $\sigma \in I$, we have $x\pi[0,\sigma s] \cap N_2 = \emptyset$, and thus $h(q(x),\sigma) = q(x\pi\sigma s)$.
Hence, in the case $N_1 \cap M_2 = \emptyset$, we have proved that h is well-defined and continuous. Obviously, h is also base-point preserving in this case.
To prove the proposition, we must therefore prove continuity of h if $N_1 \cap M_2 \neq \emptyset$.
Hence, assume $N_1 \cap M_2 \neq \emptyset$.
Let $(z_0,\sigma_0) \in N_1/M_2 \times I$, and U be open in N_1/M_2 such that $h(z_0,\sigma_0) \in U$. Then $q^{-1}[U]$ is open in N_1, i.e., there is a set W, open in X, such that $W \cap N_1 = q^{-1}[U]$.

We consider now various cases:

<u>1. Case</u>: $h(z_0,\sigma_0) \neq [M_2]$.

Since h is base-point preserving, it follows that $z_0 \neq [M_2]$ and so $q^{-1}(\{z_0\}) = \{x_0\}$ for some $x_0 \in N_1 \setminus M_2$. Moreover, $x_0\pi[0,\sigma_0 s] \cap N_2 = \emptyset$, $x_0\pi\sigma_0 s \in W \cap N_1$.
Since M_2 is closed, there is an open set W' (open in X), $W' \cap M_2 = \emptyset, x_0 \in W'$, and I' open in I, $\sigma_0 \in I'$, such that for $x \in W'$ and $\sigma \in I'$, $x\pi[0,\sigma s]$ is defined, $x\pi[0,\sigma s] \cap N_2 = \emptyset$, and $x\pi\sigma s \in W$. By Proposition 7.3, for every $\sigma \in I'$, $(W' \cap N_1)\pi\sigma s \subset W \cap N_1$.
Let $V = q(W' \cap N_1)$. Since $W' \cap M_2 = \emptyset$, $q^{-1}(q(W' \cap N_1)) = W' \cap N_1$, hence V is open in N_1/M_2. If $(z,\sigma) \in V \times I'$, then $z = q(x)$, $x \in W' \cap N_1$, $x\pi[0,\sigma s] \cap N_2 = \emptyset$, and $x\pi\sigma s \in W \cap N_1$. It follows that $h(z,\sigma) = q(x\pi\sigma s) \in q(W \cap N_1) \subset U$.

This proves continuity of h at (z_0,σ_0).

<u>2. Case:</u> $h(z_0,\sigma_0) = [M_2]$:

It follows that $N_1 \cap M_2 \subset W$.

<u>2.1 Case</u>: $z_0 \neq [M_2]$ and N_1 is N-positively invariant.

Define
$\Omega=\{(x,\sigma)\in X\times I \mid \sigma s<\omega_x$ and $x\pi\sigma s\in W$, or else $x\pi t\notin\tilde{N}_1$ for some $t<\omega_x$, $t\in[0,\sigma s]\}$.

Since $\tilde{N}_1$ is closed, Ω is open.
Let x_0 be such that $q^{-1}(\{z_0\})=\{x_0\}$.

We will show that $(x_0,\sigma_0)\in\Omega$. This is certainly true if $x_0\pi t\notin\tilde{N}_1$ for some $t<\omega_{x_0}$, $t\in[0,\sigma_0 s]$. Hence, assume that $x_0\pi t\in\tilde{N}_1$ for all $t<\omega_x$, $t\in[0,\sigma_0 s]$. Since π does not explode in N, it follows that $x_0\pi[0,\sigma_0 s]$ is defined and $x_0\pi[0,\sigma_0 s]\subset\tilde{N}_1\subset N$.

Since $h(z_0,\sigma_0)=[M_2]$, it follows that either $(x_0\pi[0,\sigma_0 s])\cap N_2=\emptyset$, $h(z_0,\sigma_0)=q(x_0\pi\sigma_0 s)$, i.e., $x_0\pi\sigma_0 s\in N_1\cap M_2\subset W$, i.e., $(x_0,\sigma_0)\in\Omega$, or else assume $x_0\pi[0,\sigma_0 s]\cap N_2\neq\emptyset$. Since N_2 is N-positively invariant $x_0\pi\sigma_0 s\in N_2\subset M_2$, and since N_1 is N-positively invariant, $x_0\pi\sigma_0 s\in N_1\cap M_2\subset W$.

Hence $(x_0,\sigma_0)\in\Omega$, as claimed.

Choose sets W', I' open in X and I, respectively, such that $(x_0,\sigma_0)\in W'\times I'$, $W'\times I'\subset\Omega$, and $W'\cap M_2=\emptyset$. Let $V=q(W'\cap N_1)$. It follows that $q^{-1}(V)=W'\cap N_1$, hence V is open in N_1/M_2 and $z_0\in V$. Let $(z,\sigma)\in V\times I'$. If $h(z,\sigma)=[M_2]$, then $h(z,\sigma)=h(z_0,\sigma_0)\in U$, and we are done. If $h(z,\sigma)\neq[M_2]$, then $z=q(x)$, $x\in W'\cap N_1$, $x\pi[0,\sigma s]$ is defined and $x\pi[0,\sigma s]\cap N_2=\emptyset$.

Since $N_1\setminus N_2\subset\tilde{N}_1$, and $\langle\tilde{N}_1,N_2\rangle$ is an index pair in N, it follows that $x\pi[0,\sigma s]\subset\tilde{N}_1$. But as $(x,\sigma)\in\Omega$, this means that $x\pi\sigma s\in W$. Also, $x\pi[0,\sigma s]\subset N_1$ since N_1 is $X\setminus N_2$-positively invariant. Hence $x\pi\sigma s\in W\cap N_1$, i.e., $h(z,\sigma)=q(x\pi\sigma s)\in U$. This proves continuity of h at (z_0,σ_0).

<u>2.2 Case</u>: $z_0\neq[M_2]$, $N_1=M_1^{\tilde{t}}$, M_1 and $\tilde{t}$ as in Definition 7.2. As before, $z_0=q(x_0)$, $x_0\in N_1\setminus M_2$.
Let

$\Omega=\{(x,\sigma)\in X\times I \mid \sigma s<\omega_x$ and $x\pi\sigma s\in W$ or else $x\pi t\notin M_1$ for some $t<\omega_x, t\in[0,\sigma s]\}$.

Since M_1 is closed, Ω is open. We will prove that $(x_0,\sigma_0)\in\Omega$. This is of course true if $x_0\pi t\notin M_1$ for some $t<\omega_{x_0}$, $t\in[0,\sigma_0 s]$. Hence assume that $x_0\pi t\in M_1$ for all $t<\omega_{x_0}$, $t\in[0,\sigma_0 s]$.

Since $M_1=(M_1\setminus N_2)\cup(M_1\cap N_2)\subset\tilde{N}_1\cup N_2\subset N$ and π does not explode in N, it follows that $x_0\pi[0,\sigma_0 s]$ is defined and $x_0\pi[0,\sigma_0 s]\subset M_1$.

Since $N_1=M_1^{\tilde{t}}$, $x_0\pi[0,\sigma_0 s]\subset N_1$. Moreover, since $h(z_0,\sigma_0)=[M_2]$, it follows, as before in 2.1 Case, that $x_0\pi\sigma_0 s\in M_2$. Hence again, $x_0\pi\sigma_0 \in N_1\cap M_2\subset W$. Now choose W' and I' with the same properties as in 2.1 Case.
Define $V=q(W'\cap N_1)$.
As before $z_0\in V$ and V is open.
Let $(z,\sigma)\in V\times I$. If $h(z,\sigma)=[M_2]$, then we are done, again. If $h(z,\sigma)\neq[M_2]$, then $z=q(x)$, $x\in W'\cap N_1$, $x\pi[0,\sigma s]$ is defined and $x\pi[0,\sigma s]\cap N_2=\emptyset$. As before, it follows that $x\pi[0,\sigma s]\subset\tilde{N}_1\setminus N_2\subset N\setminus N_2$. Moreover, $x\in N_1=M_1^{\tilde{t}}\subset M_1$. Since M_1 is $N\setminus N_2$-positively invariant, $x\pi[0,\sigma s]\subset M_1$. Hence $x\pi[0,\sigma s]\subset N_1$.

On the other hand, since $(x,\sigma)\in\Omega$, it also follows that $x\pi\sigma s\in W$, hence $x\pi\sigma s\in W\cap N_1$, i.e., $h(z,\sigma)=q(x\pi\sigma s)\in U$. Again, it follows that h is continuous at (z_0,σ_0).

<u>2.3 Case</u>: $z_0=[M_2]$ and N_1 is N-positively invariant.
Define
$W'=\{x\in X\ |\text{for all } \sigma\in I\text{: either } \sigma s<\omega_x \text{ and } x\pi[0,\sigma s]\subset W \text{ or else } x\pi t\notin\tilde{N}_1$ for some $t<\omega_x$, $t\in[0,\sigma s]\}$.

We will show that W' is open: let $x\in W'$. If $x\pi[0,s]$ is defined and $x\pi[0,s]\subset W$, or if $x\notin\tilde{N}_1$, then the same is true in a neighborhood of x. Hence, in this case, W' contains a neighborhood of x.
Therefore assume that $x\in\tilde{N}_1$ and either $s\geq\omega_x$ or $x\pi[0,s]\not\subset W$.
Let $\sigma_1=\sup\{\sigma\in I\,|\,\sigma s<\omega_x \text{ and } x\pi[0,\sigma s]\subset W\}$. Since $x\in\tilde{N}_1\cap W'$, $x\in W$, therefore $\sigma_1>0$. Moreover, $x\pi[0,\sigma_1 s)$ is defined and $x\pi[0,\sigma_1 s)\subset W$. If we assume that $x\pi[0,\sigma_1 s)\subset\tilde{N}_1\subset N$, then, from the hypothesis that π does not explode in N, we get that $x\pi\sigma_1 s$ is defined, and $x\pi[0,\sigma_1 s]\subset\tilde{N}_1$. Moreover, $x\pi\sigma_1 s\notin W$. However, this yields a contradiction to the definition of W'. Therefore, there is a $0<\sigma_2<\sigma_1$, such that $x\pi\sigma_2 s\notin\tilde{N}_1$.

Hence there is an open set $\tilde{W}$ (open in X), $x\in\tilde{W}$ such that $\tilde{W}\pi[0,\sigma_2 s]$ is defined, $\tilde{W}\pi[0,\sigma_2 s]\subset W$ and $(\tilde{W}\pi\sigma_2 s)\cap\tilde{N}_1=\emptyset$. Let $y\in\tilde{W},\sigma\in I$. Then, if $\sigma\leq\sigma_2$, it follows that $y\pi[0,\sigma s]\subset W$. If $\sigma>\sigma_2$, then, setting $t=\sigma_2 s$, it follows that $t<\omega_y$, $t\in[0,\sigma s]$, $y\pi t\notin\tilde{N}_1$.

Consequently $y\in W'$. It follows that W' is open. Moreover, if $x\in N_1\cap M_2$ and $\sigma\in I$, then either $x\pi t\notin\tilde{N}_1$ for some $t<\omega_x$, $t\in[0,\sigma s]$, or else it follows that $x\pi[0,\sigma s]$ is defined and $x\pi[0,\sigma s]\subset\tilde{N}_1$. Therefore in this case, the fact that N_1 and M_2 are N-positively invariant, implies that $x\pi[0,\sigma s]\subset N_1\cap M_2\subset W$. This implies $N_1\cap M_2\subset W'$.
Let $V=q(W'\cap N_1)$. It follows that $q^{-1}(V)=W'\cap N_1$, hence V is open in N_1/M_2. Also, $z_0\in V$. Let $z\in V$ and $\sigma\in I$. Let $x\in q^{-1}(\{z\})$. Then $x\in W'\cap N_1$. If $h(z,\sigma)\neq[M_2]$, then $x\pi[0,\sigma s]$ is defined, $x\pi[0,\sigma s]\cap N_2=\emptyset$, and $h(z,\sigma)=q(x\pi\sigma s)$. It follows that $x\pi[0,\sigma s]\subset N_1\setminus N_2\subset\tilde{N}_1$, and since $x\in W'$

this implies $x\pi\sigma s \subset W \cap N_1$, i.e., $h(z,\sigma) \in U$. This proves the continuity of h in this case.

2.4 Case: $z_0=[M_2]$, $N_1=M_1^{\tilde{t}}$, M_1 and $\tilde{t}$ as in Definition 7.2.

Define

$W'=\{x \in X \mid$ for all $\sigma \in I$: either $\sigma s<\omega_x$ and $x\pi[0,\sigma s]\subset W$ or else $x\pi t \notin M_1$ for some $t<\omega_x$, $t \in [0,\sigma s]\}$.

The same argument as in 2.3 Case (replacing $\tilde{N}_1$ by M_1) proves that W' is open. Moreover, let $x \in N_1 \cap M_2$ and $\sigma \in I$. Either $x\pi t \notin M_1$ for some $t<\omega_x$, $t \in [0,\sigma s]$, or else it follows that $\sigma s<\omega_x$ and $x\pi[0,\sigma s]\subset M_1$. In this latter case, $N_1=M_1^{\tilde{t}}$ implies that $x\pi[0,\sigma s]\subset M_1^{\tilde{t}}=N_1$ and, since M_2 is N-positively invariant, it also follows that $x\pi[0,\sigma s]\subset M_2$. Hence $x\pi[0,\sigma s]\subset N_1 \cap M_2 \subset W$. It follows that $N_1 \cap M_2 \subset W'$.

Let $V=q(W' \cap N_1)$. Then $q^{-1}(V)=W' \cap N_1$, i.e., V is open and $z_0 \in V$. Let $z \in V$ and $\sigma \in I$.

Let $x \in q^{-1}(\{z\})$. If $h(z,\sigma)=[M_2]$, then $h(z,\sigma) \in U$ and we are done. If $h(z,\sigma)\neq[M_2]$, then $\sigma x<\omega_x$, $x\pi[0,\sigma s]\cap N_2=\emptyset$, and $h(z,\sigma)=q(x\pi\sigma s)$. Since $x \in W' \cap N_1$, it follows that $x\pi[0,\sigma s]\subset N_1 \setminus N_2 \subset M_1$, and therefore $x\pi\sigma s \in W \cap N_1$, i.e., $h(z,\sigma) \in U$.

This proves continuity of h in this last case and the proof is complete.

Proposition 8.2

Let N be closed in X such that π does not explode in X, and let $\langle N_1,N_2\rangle$ be a quasi-index pair in N. Let $s\geq 0$.

Define

$$g: N_1/N_2^{-s} \to N_1^s/N_2$$

by

$$g(z) = \begin{cases} \tilde{q}(x\pi s), & \text{if there is an } x \in N_1,\ q(x)=z,\ x\pi[0,s] \\ & \text{is defined and } x\pi[0,s]\cap N_2=\emptyset . \\ [N_2], & \text{otherwise.} \end{cases}$$

Here $q: N_1 \to N_1/N_2^{-s}$, $\tilde{q}: N_1^s \to N_1^s/N_2$ are the quotient maps. Then g is a well-defined, continuous and base-point preserving mapping.

Proof:

Let $\langle\tilde{N}_1,N_2\rangle$ be a corresponding index pair. Define $G: N_1 \to N_1^s/N_2$ as

$$G(x) = \begin{cases} \tilde{q}(x\pi s), & \text{if } x \in N_1,\ x\pi[0,s] \text{ is defined and} \\ & x\pi[0,s] \cap N_2 = \emptyset\ , \\ [N_2], & \text{otherwise.} \end{cases}$$

Let $x \in N_1 \cap N_2^{-s}$. Then there is a $0 \leq t \leq s$, $t<\omega_x$, such that $x\pi[0,t] \subset N$ and $x\pi t \in N_2$. By the definition of G, $G(x)=[N_2]$.
This implies that there is a unique base-point preserving mapping $\tilde{g}: N_1/N_2^{-s} \to N_1^s/N_2$ such that $G=\tilde{g} \circ q$.
Obviously, $\tilde{g}=g$, hence g is uniquely determined and base-point preserving.
Moreover, since N_1/N_2^{-s} carries the terminal topology with respect to g, it follows that g is continuous if and only if G is. Therefore it suffices to prove that G is continuous.
Let $x_0 \in N_1$. Let U be open in N_1^s/N_2 with $G(x_0) \in U$. Then there is an open set W in X such that $N_1^s \cap W = \tilde{q}^{-1}(U)$.

<u>1. Case</u>: $G(x_0) \neq [N_2]$.

Then $s<\omega_{x_0}$, $x_0\pi[0,s] \cap N_2 = \emptyset$, and $G(x_0)=\tilde{q}(x_0\pi s)$. There is an open set V in X, $x_0 \in V$, such that for all $x \in V$, $s<\omega_x$, $x\pi[0,s] \cap N_2=\emptyset$, and $x\pi s \in W$. Then $V \cap N_1$ is open in N_1, $x_0 \in V \cap N_1$.

Let $x \in V \cap N_1$. Then $x\pi[0,s] \cap N_2 = \emptyset$. Since N_1 is $X \setminus N_2$-positively invariant, $x\pi[0,s] \subset N_1 \setminus N_2$. Moreover, $x\pi s \in W$, and $G(x)=\tilde{q}(x\pi s)$, so $G(x) \in U$.

<u>2. Case</u>: $G(x_0)=[N_2]$. Then $N_1^s \cap N_2 \subset W$.

<u>2.1 Case</u>: N_1 is N-positively invariant:

Define
$W' = \{x \in X \mid s<\omega_x$ and $x\pi s \in W$ or else there is a $t \leq s$, $t<\omega_x$, such that $x\pi t \notin \tilde{N}_1\}$.

Since $\tilde{N}_1$ is closed, W' is open in X. We prove that $x_0 \in W' \cap N_1$. In fact, either there is a $t \leq s$, $t<\omega_x$, with $x_0\pi t \notin \tilde{N}_1$, and so $x_0 \in W'$, or else suppose that $x_0\pi t \in \tilde{N}_1$ for all $t \leq s$, $t<\omega_{x_0}$. Since $\tilde{N}_1 \subset N$ it follows that $s<\omega_{x_0}$ and $x_0\pi[0,s] \subset \tilde{N}_1$. Since $G(x_0)=[N_2]$, it follows that $x_0\pi[0,s] \cap N_2 \neq \emptyset$. But since N_1 and N_2 are N-positively invariant, $x_0\pi[0,s] \subset N_1$ and $x_0\pi s \in N_2$. Hence $x_0\pi s \in N_1^s \cap N_2 \subset W$, and so $x_0 \in W'$. If $x \in W' \cap N$ and $G(x)=[N_2] \in U$, then we are done. Therefore suppose $G(x) \neq [N_2]$. Then $x\pi[0,s]$ is defined, $x\pi[0,s] \cap N_2 = 0$. It follows that $x\pi[0,s] \subset \tilde{N}_1$, but since $x \in W' \cap N_1$, we have $x\pi s \in W$ and $G(x)=\tilde{q}(x\pi s) \in U$.
Consequently $G(W' \cap N_1) \subset U$, which proves continuity at x_0.

<u>2.2 Case</u>: There is a set M_1 and $\tilde{t}$ as in Definition 7.2. Let $W'=\{x\in X | s<\omega_x,\ x\pi s\in W$ or else there is a $t\leq s$, $t<\omega_x$ with $x\pi t\notin M_1\}$.
Since M_1 is closed, W' is open.
We prove that $x_0\in W'\cap N_1$. In fact, if $x_0\pi t\notin M_1$ for some $t\leq s$, $t<\omega_{x_0}$, we are done. Suppose this is not true. Since $M_1\subset N$, $s<\omega_{x_0}$ and $x\pi[0,s]\subset M_1$.
Since $M^{\tilde{t}}=N_1$, therefore $x_0\pi[0,s]\subset N_1$. We cannot have $x_0\pi[0,s]\cap N_2=\emptyset$ since $G(x_0)=[N_2]$. Therefore, as before $x_0\pi s\in N_2\cap N_1^s\subset W$. Hence, indeed, $x_0\in W'\cap N_1$.
Let $x\in W'\cap N_1$. If $G(x)=[N_2]\in U$, we are done. Assume $G(x)\neq[N_2]$. Then $s<\omega_x$, $x\pi[0,s]\cap N_2=\emptyset$.
Since $M_1^{\tilde{t}}=N_1$ and M_1 is $N\setminus N_2$-positively invariant, we have $x\pi[0,s]\subset M_1$.
Since $x\in W'\cap N_1$, we get $x\pi s\in W\cap N_1^s$ and $G(x)=\tilde{q}(x\pi s)\in U$.
This proves the continuity of G and completes the proof of the proposition.

1.9 The Categorial Morse Index

In the preceding sections we introduced and analyzed index and quasi-index pairs. As we said before, an isolating block B determines a special index pair $\langle B,B^-\rangle$ in B. We may ask the following, more general question: given a strongly π-admissible isolating neighborhood N of K, does there exist an index pair in N? The answer is positive, under an additional assumption:

<u>Theorem 9.1</u>

<u>If N is a strongly π-admissible neighborhood of an isolated, π-invariant set K, and if there is another strongly π-admissible isolating neighborhood $\tilde{N}$ of K, such that $N\subset \mathrm{Int}\,\tilde{N}$, then for every open set W with $K\subset W$, there exists an index pair $\langle N_1,N_2\rangle$ in N with $\mathrm{Cl}(N_1\setminus N_2)\subset W$.</u>

<u>Proof</u>:
If $K=\emptyset$, let $N_1=N_2=N$. Then $\langle N_1,N_2\rangle$ is obviously an index pair with the desired properties.
Suppose now that $K\neq\emptyset$. Let $V,U,\tilde{U}$ be open sets such that $K\subset V$, $\mathrm{Cl}V\subset U\cap W$, $\mathrm{Cl}U\subset N\subset\tilde{U}$ and $\mathrm{Cl}\tilde{U}\subset\tilde{N}$. V exists since K is compact and X is normal. By making $\tilde{N}$ smaller, if necessary, we may assume that $\mathrm{Cl}\tilde{U}=\tilde{N}$. Moreover, suppose that π is a global semiflow. Since $\tilde{N}$ is strongly π-admissible, the reader will find that the proof of the general case is only notationally more complicated.
Define the functions $\tilde{g}^+$ and $\tilde{g}^-$ exactly like g^+ g^- in (i)-(vi) in the proof of Theorem 5.1 but with N,U replaced by $\tilde{N},\tilde{U}$.

Proceeding exactly as in the proof of Lemma 5.4 we find that there is an $\varepsilon>0$ such that

$$\mathrm{Cl}\{x\in\tilde{U}\mid \tilde{g}^{+}(x)<\varepsilon \text{ and } \tilde{g}^{-}(x)<\varepsilon\}\subset V .$$

Define

$$N_1 = N\cap \mathrm{Cl}\{x\in\tilde{U}\mid \tilde{g}^{-}(x)<\varepsilon\}$$

$$N_2 = N\setminus\{x\in\tilde{U}\mid \tilde{g}^{+}(x)<\varepsilon\}$$

It follows that $N_1,N_2\subset N$, N_1 is closed, and N_2 is closed since $\tilde{g}^{+}$ is upper-semicontinuous by Proposition 5.2.

Let $y\in N_1$, $y\pi[0,s]\subset N$. Then there are $x_n\in\tilde{U}$, $x_n\to y$ as $n\to\infty$, and $\tilde{g}^{-}(x_n)<\varepsilon$. Since $N\subset\tilde{U}$ and $\tilde{U}$ is open, $x_n\pi[0,s]$ is defined for large n. Moreover, for all $0\leq t\leq s$, $\tilde{g}^{-}(x_n\pi t)\leq\tilde{g}^{-}(x_n)<\varepsilon$ by Proposition 5.2. This implies that $y\pi t\in N_1$ for $0\leq t\leq s$, and so N_1 is N-positively invariant.

Now let $y\in N_2$, $y\pi[0,s]\subset N$. Then for $0\leq t\leq s$, $\tilde{g}^{+}(y\pi t)$ is defined and $\tilde{g}^{+}(y\pi t)\geq\tilde{g}^{+}(y)\geq\varepsilon$, i.e. $y\pi t\in N_2$. Hence N_2 is N-positively invariant. Since $\tilde{g}^{+}$ and $\tilde{g}^{-}$ are both upper-semicontinuous we get that $\tilde{W}:=\{x\in\tilde{U}\mid\tilde{g}^{+}(x)<\varepsilon \text{ and } \tilde{g}^{-}(x)<\varepsilon\}$ is open, $K\subset\tilde{W}$, $\tilde{W}\subset N_1\setminus N_2$.
It follows that $K\subset\mathrm{Int}(N_1\setminus N_2)$.
Let $x\in N_1$ be such that for some $t>0$, $x\pi t\notin N$. Let $t'=\sup\{s\mid x\pi[0,s]\subset N\}$. Since N is closed, $x\pi[0,t']\subset N$ and by the N-positive invariance of N_1, $x\pi[0,t']\subset N_1$. Suppose that $x\pi t'\notin N_2$.
Then $\tilde{g}^{+}(x\pi t')<\varepsilon$ and it follows that $x\pi t'\in\mathrm{Cl}\tilde{W}$. But $\mathrm{Cl}\tilde{W}\subset V$ by our hypothesis. Hence $x\pi t'\in V$, which contradicts the definition of t'.
All this proves that $\langle N_1,N_2\rangle$ is an index pair in N. It remains to be shown that

$$\mathrm{Cl}(N_1\setminus N_2)\subset W.$$

Indeed, $N_1\setminus N_2\subset V$ and $\mathrm{Cl}V\subset W$, by our assumptions. The proof is complete.

Given an isolated π-invariant set K and a strongly π-admissible isolating neighborhood N of K, there is, as we have just proved, an index pair $\langle N_1,N_2\rangle$ in N. By the results of Section 1.7, for every $s,r\geq 0$, $\langle N_1^{s},N_1^{-r}(N)\rangle$ will be a quasi-index pair in N.
Furthermore, different choices of isolating neighborhoods of K will lead to yet different index and quasi-index pairs. Hence the question arises if all these different quasi-index pairs for K are somehow re-

lated to each other. The answer is yes and it leads to the concept of the (categorial) Morse index defined below.

For the rest of this section, let X be a metric space and π be a local semiflow on X. If $Y \subset X$, then $s_Y: Y \to \mathbb{R}$ is defined as $s_Y(x) = \sup\{t \geq 0 \mid t < \omega_x,\ x\pi[0,t] \subset Y\}$.

Recall that T* is the category of pointed spaces and continuous basepoint preserving maps, and HT* is the corresponding homotopy category.

Definition 9.2

Let K be a compact, isolated π-invariant set, and $N = N(\pi,K)$ be the set of all isolating strongly π-admissible neighborhoods N of K. Assume that $N \neq \emptyset$. Let $V = V(\pi,K)$ be the smallest subcategory of T* such that

(1) For every $N \in N$ and every quasi-index pair $\langle N_1,N_2\rangle$ in N, $(N_1/N_2,[N_2])$ is an object of V.

(2) If $N,\bar{N} \in N$ and $\langle N_1,N_2\rangle$ and $\langle \bar{N}_1,\bar{N}_2\rangle$ are quasi-index pairs in N and $\bar{N}$, resp. such that $N_1 \setminus N_2 \subset \bar{N}_1$ and $(N_1 \cup \mathrm{Cl}(N_1 \setminus N_2)) \cap N_2 \cap \bar{N}_1 \subset \bar{N}_2$ and if the inclusion induced map i: $N_1/N_2 \to \bar{N}_1/\bar{N}_2$ is admissible, then i is a morphism of V.

Moreover, if $s \geq 0$ and g^s is defined as in Proposition 8.2 then g^s is a morphism of V.

Now the Morse index $I(\pi,K)$ is the category defined as follows:

(1) The class of objects of $I(\pi,K)$ is equal to the class of objects of V, i.e., the class of all pointed sets $(N_1/N_2,[N_2])$ where $\langle N_1,N_2\rangle$ is a quasi-index pair in N, for some $N \in N$.

(2) The morphisms in $I(\pi,K)$ are the homotopy classes [f] in HT* of morphisms f from V.

Definition 9.3

A small category C is called a connected simple system (c.s.s.) if for all pairs (A,B) of objects in C, the set hom(A,B) of morphisms from A to B contains exactly one element. This element is then necessarily an isomorphism.

The main result of this section is

Theorem 9.4

$I(\pi,K)$ is a connected simple system.

To prove Theorem 9.4 we need the following

Proposition 9.5

If $N \in N$, $\langle N_1,N_2\rangle$ is a quasi-index pair in N and $f: N_1/N_2 \to N_1/N_2$ is a morphism in V, then f is homotopic in T to the identity map I_{N_1/N_2}.*

Proof:

By Definition 9.2, we can assume that there are numbers $t_i \geq 0$, sets $M_i \in N$, quasi-index pairs $\langle M_{i,1}, M_{i,2}\rangle$ in M_i, and mappings j_0, j_i, g_i, $i=1,\ldots,n$ such that

(1) $g_i = g^{t_i}$, where g^{t_i} is defined in Proposition 8.2 (with $s=t_i$ and $\langle N_1,N_2\rangle = \langle M_{i,1}, M_{i,2}\rangle$ in that proposition).

(2) j_0 and j_i are admissible inclusion induced mappings.

(3) $f = j_n \circ g_n \circ \ldots \circ j_1 \circ g_1 \circ j_0$.

These conditions imply that

$$g_i : M_{i,1}/M_{i,2}^{-t_i} \to M_{i,1}^{t_i}/M_{i,2}, N_1 \setminus N_2 \subset M_{1,1} ,$$

$$(N_1 \cup Cl(N_1 \setminus N_2)) \cap N_2 \cap M_{1,1} \subset M_{1,2}^{-t_i}, M_{i,1}^{t_i} \setminus M_{i,2} \subset M_{i+1,1},$$

$$(M_{i,1}^{t_i} \cup Cl(M_{i,1}^{t_i} \setminus M_{i,2})) \cap M_{i,2} \cap M_{i+1,1} \subset M_{i+1,2}^{-t_{i+1}},\ i=1,\ldots,n \text{ , where}$$

$$\langle M_{n+1,1}, M_{n+1,2}^{-t_{n+1}}\rangle = \langle N_1, N_2\rangle.$$

We shall need the following lemmas:

Lemma 9.6

Let M, $\overline{M} \in N$, *and let* $\langle M_1, M_2\rangle$ *and* $\langle \overline{M}_1, \overline{M}_2\rangle$ *be two quasi-index pairs in* M *and* $\overline{M}$, *respectively. Assume that* $M_1 \setminus M_2 \in \overline{M}_1$ *and* $(M_1 \cup Cl(M_1 \setminus M_2)) \cap M_2 \cap \overline{M}_1 \subset M_2$. *Then*

(1) *For every* $s \geq 0$
$(M_1 \setminus M_2) \cap (\overline{M}_1 \setminus \overline{M}_2) \cap M_2^{-s} \subset \overline{M}_2^{-s}$.

(2) *For some* $T \geq 0$
$A_T(M_1) := \{x \mid x\pi[0,T] \subset M_1\} \subset \overline{M}_1 \setminus \overline{M}_2$.

Proof:

(1) Suppose that for some $s \geq 0$ and $x \in X, x \in (M_1 \setminus M_2) \cap (\overline{M}_1 \setminus \overline{M}_2) \cap M_2^{-s} \cap (X \setminus \overline{M}_2^{-s})$. Since $\overline{M}_1$ is $X \setminus \overline{M}_2$-positively invariant (by Proposition 7.3), it follows that $x\pi[0,s] \subset \overline{M}_1 \setminus \overline{M}_2$. Let $t_0 = \sup\{t \mid x\pi[0,t] \cap M_2 = \emptyset\}$. Then $t_0 > 0$ and $t_0 \leq s$. Also, $x\pi[0,t_0) \subset M_1 \setminus M_2$. Hence

$$x\pi t_0 \in Cl(M_1 \setminus M_2) \cap \overline{M}_1 \cap M_2 \subset \overline{M}_2 ,$$

a contradiction which proves (1).

(2) Let $s = \sup\{s_M(x) \mid x \in M_2\}$, $s = \sup\{s_M(x) \mid x \in M_2\}$. Since M and $\overline{M}$ are strongly π-admissible, it follows that $s + \overline{s} < \infty$. Let $\delta > 0$ and $T > s + \overline{s} + 2\delta$.

Choose $x \in A_T(M_1)$ arbitrarily. Then $x\pi[0,T] \subset M_1$; thus $x\pi[0,T-s-\delta] \subset M_1 \setminus M_2 \subset \bar{M}_1$, and this in turn implies that

$$x\pi[0,(T-s-\delta)-\bar{s}-\delta] \subset \bar{M}_1 \setminus \bar{M}_2$$

This implies $x \in \bar{M}_1 \setminus \bar{M}_2$.

<u>Lemma 9.7</u>

<u>There is a</u> $T>0$ <u>such that for every</u> $k=1,\ldots,n+1$

$$A_T(N_1)\pi\left(\sum_{i=1}^{k-1} t_i\right) \subset M_{k,1} \setminus M_{k,2}^{-t_k} .$$

<u>Here</u>, $\sum_{i=1}^{0} := 0$.

<u>Proof</u>:

We shall show that for every $k=1,\ldots,n+1$, there is a T_k, such that

$$A_{T_k}(N_1)\pi\left(\sum_{i=1}^{k-1} t_i\right) \subset M_{k,1} \setminus M_{k,2}^{-t_k} . \tag{1}$$

Inclusion (1) is true for $k=1$, by Lemma 9.6. Suppose that (1) holds for all $1 \le k \le m$. In particular, for some T_m

$$A_{T_m}(N_1)\pi\left(\sum_{i=1}^{m-1} t_i\right) \subset M_{m,1} \setminus M_{m,2}^{-t_m} .$$

Consequently

$$A_{T_m}(N_1)\pi\left(\sum_{i=1}^{m-1} t_i\right)\pi[0,t_m] \cap M_{m,2} = \emptyset .$$

Since $M_{m,1}$ is $X \setminus M_{m,2}$-positively invariant it follows that

$$A_{T_m}(N_1)\pi\left(\sum_{i=1}^{m} t_i\right) \subset M_{m,1}^{t_m} \setminus M_{m,2} .$$

By Lemma 9.6, there is a $T' \ge T_m$ such that

$$A_{T'}(M_{m,1}^{t_m}) \subset M_{m+1,1} \setminus M_{m+1,2}^{-t_{m+1}} .$$

Let $T_{m+1} = 2T'$. We will show that

$$A_{T_{m+1}}(N_1)\pi\left(\sum_{i=1}^{m} t_i\right) \subset A_{T'}(M_{m,1}^{t_m}) ,$$

and this will prove the lemma. In fact, let $x \in A_{T_{m+1}}(N_1)\pi(\sum_{i=1}^{m} t_i)$.

Then there is a $y \in A_{T_{m+1}}(N_1)$ such that $y\pi(\sum_{i=1}^{m} t_i)=x$. Let $s \in [0,T']$. Then $(y\pi s)\pi[0,T_m]\subset y\pi[0,2T']\subset N_1$. Hence $y\pi s \in A_{T_m}(N_1)$, i.e.,

$$x\pi s \in A_{T_m}(N_1)\pi\left(\sum_{i=1}^{m} t_i\right)\subset M_{m,1}^{t_m}\setminus M_{m,2} .$$

This implies that $x\pi[0,T']\subset M_{m,1}^{t_m}$, i.e., $x \in A_{T'}(M_{m,1}^{t_m})$.

Lemma 9.8

For $i=1,\ldots,n+1$, let p_i be the base point of $M_{i,1}/M_{i,2}^{-t_i}$ and $q_i:M_{i,1}\to M_{i,1}/M_{i,2}^{-t_i}$ the quotient map. (Remember that $\langle M_{n+1,1},M_{n+1,2}^{-t_{n+1}}\rangle=\langle N_1,N_2\rangle$.) Let $\bar{t}=\sum_{i=1}^{n} t_i$. Then whenever $y \in N_1\cap N_2^{-\bar{t}}$,

$$q_{n+1}^{-1}\{fq_{n+1}(y)\}\subset N_1\cap N_2^{-\bar{t}} .$$

Proof:

Let $f_k=j_k\circ g_k\circ\ldots\circ j_1\circ g_1\circ j_0$, $j=1,\ldots,n$. Let $y \in N_1\cap N_2^{-\bar{t}}$. If $fq_{n+1}(y)=p_{n+1}$ then

$$q_{n+1}^{-1}\{fq_{n+1}(y)\} = N_1\cap N_2, \text{ if } N_1\cap N_2\neq\emptyset ,$$

$$= \emptyset \quad \text{otherwise.}$$

It follows that $q_{n+1}^{-1}\{fq_{n+1}(y)\}\subset N_1\cap N_2\subset N_1\cap N_2^{-\bar{t}}$ in this case. Now assume that $fq_{n+1}(y)\neq p_{n+1}$. Then for every $k=1,\ldots,n$, $f_k q_{n+1}(y)\neq p_{k+1}$. Hence

$$f_k q_{n+1}(y) = q_{k+1}\left(y\pi\left(\sum_{i=1}^{k} t_i\right)\right), \quad \text{for } k=1,\ldots,n.$$

By our assumption, $y \notin N_2$; hence $y \in (N_1\setminus N_2)\cap N_2^{-\bar{t}}$ and $y \notin M_{1,2}^{-t_1}$. Lemma 9.6 implies

$$y \in (M_{1,1}\setminus M_{1,2}^{-t_1})\cap(M_{1,2}^{-t_1})^{-\bar{t}} ,$$

and so

$$y\pi t_1 \in (M_{1,1}^{t_1}\setminus M_{1,2})\cap M_{1,2}^{-\bar{t}} .$$

For induction assume that

$$y\pi\left(\sum_{i=1}^{k} t_i\right) \in (M_{k,1}^{t_k} \setminus M_{k,2}) \cap M_{k,2}^{-\bar{t}}\ , \qquad k \leq n-1$$

Since $f_{k+1}q_{n+1}(y) \neq p_{k+2}$, it follows that $y\pi(\sum_{i=1}^{k} t_i) \in M_{k+1,1} \setminus M_{k+1,2}^{-t_{k+1}}$. Lemma 9.6 again implies

$$y\pi\left(\sum_{i=1}^{k} t_i\right) \in (M_{k+1,1} \setminus M_{k+1,2}^{-t_{k+1}}) \cap (M_{k+1,2}^{-t_{k+1}})^{-\bar{t}}\ ;$$

hence

$$y\pi\left(\sum_{i=1}^{k+1} t_i\right) \in (M_{k+1,1}^{t_{k+1}} \setminus M_{k+1,2}) \cap M_{k+1,2}^{-\bar{t}}\ .$$

This proves the lemma by induction.

Lemma 9.9

Define $f^{\bar{t}}: N_1/N_2 \to N_1/N_2$ as $f^{\bar{t}}(z) = h(z,1)$, where h is defined in Proposition 8.1 with $M_2 = N_2$ and $s = \bar{t}$ in that proposition.

Then there is an $r \geq \bar{t}$ such that $f^{\bar{t}}q_{n+1}(y) \neq fq_{n+1}(y)$ implies

$$q_{n+1}^{-1}\{f^{\bar{t}}q_{n+1}(y), fq_{n+1}(y)\} \subset N_1 \cap N_2^{-r}\ .$$

Proof:

Let $T \geq \bar{t}$ be as in Lemma 9.7. Set $r = T$. Either $fq_{n+1}(y) = q_{n+1}(y\pi\bar{t})$ or $fq_{n+1}(y) = p_{n+1}$. Also, either $f^{\bar{t}}q_{n+1}(y) = q_{n+1}(y\pi\bar{t})$ or $f^{\bar{t}}q_{n+1}(y) = p_{n+1}$. Hence there are two possibilities:

First assume

$$fq_{n+1}(y) = p_{n+1} \neq f^{\bar{t}}q_{n+1}(y)\ .$$

Since N_1 is $X \setminus N_2$-positively invariant it follows that $N_1 \setminus N_2^{-r} \subset A_r(N_1)$. Since $fq_{n+1}(y) \neq f^{\bar{t}}q_{n+1}(y)$, it follows that $y\pi[0,\bar{t}] \in N_1 \setminus N_2$. Since $fq_{n+1}(y) = p_{n+1}$, it follows from Lemma 9.7 that $y \in N_1 \setminus A_r(N_1)$. Hence $y \in N_1 \cap N_2^{-r}$, and $y\pi\bar{t} \in N_1 \cap N_2^{-r}$. Since $q_{n+1}^{-1}\{f^{\bar{t}}q_{n+1}(y),\ fq_{n+1}(y)\} \subset \{y\pi\bar{t}\} \cup (N_1 \cap N_2) \subset N_1 \cap N_2^{-r}$, the conclusion follows.

Now assume that

$$fq_{n+1}(y) \neq p_{n+1} = f^{\bar{t}}q_{n+1}(y)\ .$$

Hence $y \in N_1 \cap N_2^{-\bar{t}}$; now by Lemma 9.8,

$$q_{n+1}^{-1}\{fq_{n+1}(y)\} \subset N_1 \cap N_2^{-\bar{t}} \subset N_1 \cap N_2^{-r} .$$

Now the conclusion follows again and the lemma is proved. We can now complete the proof of Proposition 9.5.

Let $\rho : N_1/N_2 \to N_1/N_2^{-r}$ and $e : N_1^r/N_2 \to N_1/N_2$ be the inclusion induced mappings. Lemma 9.9 implies that

$$\rho o f^{\bar{t}} = \rho o f .$$

We write $f \sim g$ if f is homotopic to g (in T*). It follows that

$$f = I_{N_1/N_2} o f \sim f^r o f = (e o g^r o \rho) o f = (e o g^r) o (\rho o f)$$
$$= (e o g^r) o (\rho o f^{\bar{t}}) = (e o g^r o \rho) o f^{\bar{t}} = f^r o f^{\bar{t}} \sim I_{N_1/N_2} .$$

The proposition is proved.

To prove Theorem 9.4 we need two more lemmas.

Lemma 9.10

Let $M,N \in N(\pi,K)$, $M \subset N$, and suppose that $\langle M_1,M_2\rangle$ and $\langle N_1,N_2\rangle$ are quasi-index pairs in M and N, respectively. Then for all $r \geq 0$ large enough there is a $s = s(r) > r$ such that

$$M_1^r \setminus M_2^{-r} \subset N_1 \quad \text{and}$$

$$\mathrm{Cl} M_1^r \cap M_2^{-r} \cap N_1 \subset N_2^{-s} .$$

Proof:

Let $\langle \tilde{M}_1,M_2\rangle$ and $\langle \tilde{N}_1,N_2\rangle$ be the corresponding index pairs. Let $r \geq 0$, $x \in M_1^r \setminus M_2^{-r}$. Then for some $y \in X$, $y\pi[0,r] \subset M_1 \subset M$ and $x = y\pi r$. Since M_2^{-r} is M-positively invariant $y\pi[0,r] \cap M_2^{-r} = \emptyset$. Hence $y\pi[0,r] \subset M_1 \setminus M_2^{-r} \subset \tilde{M}_1$ and so $y\pi[0,r] \subset \tilde{M}_1 \setminus M_2^{-r}$. It follows that $x \in \tilde{M}_1^r \setminus M_2^{-r}$. This means that $M_1^r \setminus M_2^{-r} \subset \tilde{M}_1^r \setminus M_2^{-r}$. Therefore for all r large enough, $M_1^r \setminus M_2^{-r} \subset \tilde{M}_1^r \setminus M_2^{-r} \subset \mathrm{Int}(\tilde{N}_1 \setminus N_2)$. If s is such that $\tilde{N}_1^s \subset N_1$, it follows that $M_1^{r+s} \setminus M_2^{-r-s} \subset \tilde{N}_1^s \subset N_1$. Hence for all r large enough, $M_1^r \setminus M_2^{-r} \subset N_1$.

Now suppose that there are sequences x_n, r_n, s_n such that $x_n \in \mathrm{Cl} M_1^{r_n} \cap M_2^{-r_n} \cap N_1$, $r_n \to \infty$, $s_n \geq 3r_n$, $x_n \notin N_2^{-s_n}$. Let $y_n \in M_1^{r_n}$ be such that

$d(y_n,x_n)<2^{-n}$. Hence, for some $z_n \in X$, $z_n\pi[0,r_n]\subset M_1$, $z_n\pi r_n=y_n$. Since $<\tilde{M}_1,M_2>$ is an index pair in M it follows that $\sup\{s_M(x) \mid x \in M_2\}=r<\infty$. Hence for some t_n, $0\le t_n\le r_n+r$, it follows that $x_n\pi[0,t_n]\subset M\subset N$ and $x_n\pi t_n \in \partial M$, $x_n\pi[0,s_n]\cap N_2=\emptyset$. Taking subsequences if necessary we may assume that $r_n+r\le 2r_n$ for all n and $y_n\to y \in A^-(M)$. Consequently, $x_n\to y$ as $n\to\infty$. Taking subsequences again we may assume that $x_n\pi t_n \to w \in A^-(M)\cap\partial M$.
Moreover, since $s_n\ge 3r_n$, we get $(x_n\pi t_n)\pi[0,r_n]\subset N$. Thus $w \in A^+(N)$.
Therefore $w \in A^-(M)\cap A^+(N)\subset A^-(N)\cap A^+(N)=K$ and $w \in \partial M$, a contradiction, which proves the lemma.

Lemma 9.11

Let $N \in \mathcal{N}(\pi,K)$ and let $<N_1,N_2>$ be a quasi-index pair in N. Choose a corresponding index pair $<\tilde{N}_1,N_2>$ as in Definition 7.2. Assume that N_1 is N-positively invariant. If $M \in \mathcal{N}(\pi,K)$ and $M\subset \mathrm{Int}\tilde{N}_1$, then there is $r_0\ge 0$, such that for all $r\ge r_0$, $<N_1^r\cap M, N_2^{-r}\cap M>$ is a quasi-index pair in M.

Proof:

Choose open sets U,V such that

$$K \subset V \subset \mathrm{Cl}V \subset U \subset \mathrm{Cl}U \subset M \subset \mathrm{Int}\tilde{N}_1 .$$

As $N \in \mathcal{N}(\pi,K)$, there is an r' such that $\tilde{N}_1^{r'}\setminus N_2^{-r'}\subset V$. Now let s_0 be such that $\tilde{N}_1^{s_0}\subset N_1^{r'}$. This is possible by Definition 7.2.

Finally, let W be an open set, $K\subset W\subset V$ such that whenever $x \in W$, then $x\pi[0,s_0]\subset V$. W exists since $N \in \mathcal{N}(\pi,K)$.
Define

$$N_1' = M\cap \mathrm{Cl}\{y \in X \mid \text{there is } x \in W \text{ and } t\ge 0 \text{ such that } x\pi[0,t]\subset \mathrm{Int}\tilde{N}_1 \text{ and } x\pi t=y\}.$$

We show that $<N_1',N_2^{-r'}\cap M>$ is an index pair in M:

Obviously N_1' and $N_2^{-r'}\cap M$ are closed and $N_2^{-r'}\cap M$ is M-positively invariant. Let $x \in N_1'$, $s\ge 0$ be such that $x\pi[0,s]\subset M\subset \mathrm{Int}\tilde{N}_1$. Then there are $x_n \in W$, $t_n\ge 0$ such that $x_n\pi[0,t_n]\subset \mathrm{Int}\tilde{N}_1$ and $x_n\pi t_n\to x$. Hence for n large enough, $x_n\pi[0,t_n+s]\subset \mathrm{Int}\tilde{N}_1$. This proves that $x\pi[0,s]\subset N_1'$ and hence N_1' is M-positively invariant.

Furthermore

$$K \subset W \setminus (N_2^{-r'} \cap M) \subset \operatorname{Int}(N_1' \setminus (N_2^{-r'} \cap M)) \ .$$

Now let $x \in N_1'$ and $t' \geq 0$ be such that $x\pi t' \notin M$. Let $t_0 = s_M(x)$. Then $x\pi t_0 \in \partial M \cap N_1'$. Suppose that $x\pi t_0 \notin N_2^{-r'}$. There are $x_n \in W$, $t_n \geq 0$, $x_n\pi[0,t_n] \subset \operatorname{Int}\tilde{N}_1$, and $x_n\pi t_n \to x\pi t_0$. By the choice of V, it follows that $x_n\pi t_n \notin \operatorname{Cl}V$ for n large enough. By the choice of W, $t_n > s_0$. Hence $x_n\pi t_n \in \tilde{N}_1^{s_0} \subset N_1^{r'}$; hence for n large enough $x_n\pi t_n \in N_1^{r'} \setminus \tilde{N}_2^{-r'} \subset \tilde{N}_1^{r'} \setminus N_2^{-r'} \subset V$, a contradiction. Hence $x\pi t_0 \in N_2^{-r'}$. This proves that $\langle N_1', N_2^{-r'} \cap M\rangle$ is an index pair in M. It follows that for any $r \geq r'$, $\langle N_1', N_2^{-r} \cap M\rangle$ is an index pair in M. Choose $r_0 \geq r'$ so large that $\tilde{N}_1^{r_0} \setminus N_2^{-r_0} \subset W$. Then for $r \geq r_0$, $N_1^r \setminus N_2^{-r} \subset N_1'$ and $N_1'^{\,s_0+r-r'} \subset M \cap \tilde{N}_1^{s_0+r-r'} \subset M \cap N_1^r$. Since $N_1^r \cap M$ is M-positively invariant, it follows that $\langle N_1^r \cap M, N_2^{-r} \cap M\rangle$ is a quasi-index pair in M. We are now in a position to prove Theorem 9.4:

We claim that $\hom(A,B) \neq \emptyset$, for any two objects A,B of the category $I(\pi,K)$. Let us show that this claim implies the assertion of the theorem. In fact, let $a_1, a_2 \in \hom(A,B)$ and choose some $g \in \hom(B,A)$. Then $a_i \circ g \in \hom(B,B)$ and $g \circ a_i \in \hom(A,A)$. Proposition 9.5 implies that $a_i \circ g$ and $g \circ a_i$ are identity morphisms. This implies that $a_1 = g^{-1} = a_2$ proving that $\hom(A,B)$ contains exactly one element. Therefore it remains to prove our claim.
There are $N, \bar{N} \in \mathcal{N}$ and quasi-index pairs $\langle N_1, N_2\rangle$ respectively $\langle \bar{N}_1, \bar{N}_2\rangle$ with corresponding index pairs $\langle \tilde{N}_1, \tilde{N}_2\rangle$ respectively $\langle \tilde{\bar{N}}_1, \tilde{\bar{N}}_2\rangle$ in N respectively $\bar{N}$, such that

$$A = (N_1/N_2, [N_2]), \quad B = (\bar{N}_1/\bar{N}_2, [\bar{N}_2]).$$

Let t be such that $\tilde{N}_1^t \subset N_1$. Let $M \in \mathcal{N}$, $M \subset \operatorname{Int}(\tilde{\bar{N}}_1 \cap N)$. Lemma 9.11 implies the existence of an $r \geq 0$ such that $\tilde{\bar{N}}_1^r \setminus \bar{N}_2^{-r} \subset M$ and $\langle M_1, M_2\rangle := \langle \tilde{\bar{N}}_1^r \cap M, \bar{N}_2^{-r} \cap M\rangle$ is a quasi-index pair in M. By Proposition 6.2, there exists an admissible, inclusion induced map $e : \tilde{\bar{N}}_1^r / \bar{N}_2^{-r} \to M_1/M_2$. By Lemma 9.10, there are $r', s \geq 0$ $r' < s$ such that

$$M_1^{r'} \setminus M_2^{-r'} \subset \tilde{N}_1 \ , \quad \operatorname{Cl} M_1^{r'} \cap M_2^{-r'} \cap \tilde{N}_1 \subset N_2^{-s} \ .$$

Now, Proposition 6.2 implies again that there is an admissible inclusion induced map $j : M_1^{r'}/M_2^{-r'} \to \tilde{N}_1/N_2^{-s}$.

Consider the following sequence of morphisms in $V(\pi,K)$:

$$\bar{N}_1/\bar{N}_2 \overset{a_1}{\to} \tilde{\bar{N}}_1/\bar{N}_2 \overset{a_2}{\to} \tilde{\tilde{N}}_1/\bar{N}_2^{-r} \overset{b_1}{\to} \tilde{\bar{N}}^r/\bar{N}_2 \overset{a_3}{\to} \tilde{\bar{N}}^r/\bar{N}_2^{-r} \overset{e}{\to} M_1/M_2$$

$$\overset{a_4}{\to} M_1/M_2^{-r'} \overset{b_2}{\to} M_1^{r'}/M_2 \overset{a_5}{\to} M_1^{r'}/M_2^{-r'} \overset{j}{\to} \tilde{N}_1/N_2^{-s} \overset{b_3}{\to} \tilde{N}_1^s/N_2$$

$$\overset{a_6}{\to} \tilde{N}_1/N_2 \overset{a_7}{\to} \tilde{N}_1/N_2^{-t} \overset{b_4}{\to} \tilde{N}_1^t/N_2 \overset{a_8}{\to} N_1/N_2 \; .$$

In this sequence, all maps a_i, $i=1,\ldots,8$, are admissible inclusion induced maps, and all b_k, $k=1,\ldots,4$ are maps of the type g^s defined in Proposition 8.2.
Let f be the composition of all the maps in the sequence. Then $f:\bar{N}_1/\bar{N}_2 \to N_1/N_2$ and f is a morphism in $V(\pi,K)$. Therefore [f] is a morphism in $I(\pi,K)$, and so $\hom(A,B) \neq \emptyset$, as claimed.
The proof of Theorem 9.4 is complete.

1.10 The homotopy index and its basic properties

The following fundamental result is an immediate consequence of Theorem 9.4:

Theorem 10.1

If $N,\bar{N} \in N(\pi,K)$ and if $\langle N_1,N_2\rangle$ respectively $\langle \bar{N}_1,\bar{N}_2\rangle$ are quasi-index pairs in N respectively $\bar{N}$, then the pointed spaces $(N_1/N_2, [N_2])$ and $(\bar{N}_1/\bar{N}_2, [\bar{N}_2])$ are homotopy equivalent.
Consequently the homotopy type $[N_1/N_2, [N_2]]$ depends only on the pair (π,K). We are thus led to the following concept:

Definition 10.2

Let π be a local semiflow on the metric space X and K be an isolated invariant set relative to π which has a strongly π-admissible isolating neighborhood (relative to π). Choose any such neighborhood N and choose an arbitrary quasi-index pair $\langle N_1,N_2\rangle$ in N.
Then the *homotopy index* $h(\pi,K)$ *of* (π,K) is defined to be the homotopy type of $(N_1/N_2, [N_2])$

$$h(\pi,K) = [N_1/N_2, [N_2]] \; .$$

For two-sided flows on compact or locally compact spaces, i.e. essentially for finite-dimensional ODES, the homotopy index is due to Charles Conley (Conley [1]) and, in that case, it is also known as the *Conley index*.
In the more general case presented here, neither the two-sidedness

of the semiflow nor local compactness of X is assumed. Instead we assume admissibility, and $h(\pi,K)$ is defined whenever $N(\pi,K)\neq\emptyset$. Let S (or $S(X)$ for clarity) be the set of all pairs (π,K), where $N(\pi,K)\neq\emptyset$. In words, S is the set of all pairs (π,K), where π is a local semiflow on X, and K is an isolated invariant set relative to π for which there is an isolating neighborhood N (relative to π) which is strongly π-admissible.

In our general theory, the homotopy index (and also the categorial Morse index) is defined for all (π,K) in S. Such pairs are therefore often called <u>admissible</u>. In short, the index is defined for admissible pairs.

We will now discuss some properties of the homotopy index.

Suppose that $K=\emptyset$. Then $(\pi,\emptyset)\in S$. In fact, take $N=\emptyset$, $N_1=N_2=\emptyset$. Then $N\in N(\pi,K)$ and $\langle N_1,N_2\rangle$ is an index pair in N. Now $h(\pi,K)=[\emptyset/\emptyset,[\emptyset]]=[\{p_0\},p_0]$ where $\{p_0\}$ is any one-point space. We write $\overline{0}:=[\{p_0\},p_0]$. Therefore, by definition, $h(\pi,\emptyset)=\overline{0}$. Hence, if we know that $h(\pi,K)\neq\overline{0}$, then necessarily $K\neq\emptyset$.

Let us now discuss a trivial example. Consider a scalar ODE

$$\dot{x} = f(x) \qquad (1).$$

Let $B=[-1,1]$.

<u>1. Case</u>: $f(x)\equiv 1$.

Then $K=A(B)$, the largest invariant set in B (relative to the semiflow π generated by (10.1), is empty. Therefore $h(\pi,K)=\overline{0}$.

<u>2. Case</u>: $f(x)=x^2$.

Here, clearly, $K=A(B)=\{0\}$ and B is an isolating block with $B^-=\{1\}$. Hence $(B/B^-, [B^-])$ is contractible, i.e. $h(\pi,K)=[B/B^-, [B^-]]=\overline{0}$. Consequently, a nonempty set can have zero index.

<u>3. Case</u>: $f(x)=x$.

Now, again, $K=\{0\}$, and B is an isolating block. However, this time $B^-=\{-1,1\}$. Therefore $(B/B^-,[B^-])$ has the homotopy type of a pointed 1-sphere (S^1,s_0). Call this homotopy type $\sum^1$. Hence $h(\pi,K)=\sum^1$.

<u>4. Case</u> $f(x)=-x$.

Here, $K=\{0\}$, $B^-=\emptyset$. This means that $(B/B^-,[B^-])$ is homotopy equivalent to $(B\dot{\cup}\{p_0\})$, i.e. to $(\{p_1,p_0\}, p_0)$ where $p_1\neq p_0$. However, the latter space is a pointed 0-sphere (S^0,s_0) with its homotopy type denoted by $\sum^0$. Hence $h(\pi,K)=\sum^0$.

We will now discuss two operations for homotopy indices.

Definition 10.3

Let (Y,y_0) and (Z,z_0) be two pointed spaces.
The wedge sum or join $(Y,y_0)\vee(Z,z_0)$, frequently written $Y\vee Z$ for short, is the pointed space (W,w_0), where $W=Y\times\{z_0\}\cup\{y_0\}\times Z\subset Y\times Z$, $w_0=(y_0,z_0)$.
The smash-product $(Y,y_0)\wedge(Z,z_0)$ (also written $Y\wedge Z$) is the pointed space (W,w_0), where

$$W = (Y\times Z)/(Y\times\{z_0\}\cup\{y_0\}\times Z) \ ,$$

$$w_0 = [Y\times\{z_0\}\cup\{y_0\}\times Z].$$

It follows from Definition 10.3 that $Y\wedge Z=(Y\times Z)/(Y\vee Z)$.

It is well-known (see Spanier [1], Puppe [1]) that both the join and the smash product of (Y,y_0) and (Z,z_0) only depend on the homotopy types of (Y,y_0) and (Z,z_0) respectively. Consequently $\vee$ and $\wedge$ are well-defined on homotopy types of pointed spaces. In particular, given $(\pi_1,K_1)\in\mathcal{S}(X_1)$ and $(\pi_2,K_2)\in\mathcal{S}(X_2)$ (where X_1 and X_2 are metric spaces), it is meaningful to write $h(\pi_1,K_1)\vee h(\pi_2,K_2)$ and $h(\pi_1,K_1)\wedge h(\pi_2,K_2)$.

Now we have the following

Theorem 10.4

If $(\pi,K_1)\in\mathcal{S}(X)$, $(\pi,K_2)\in\mathcal{S}(X)$ and $K_1\cap K_2=\emptyset$, then $(\pi,K_1\cup K_2)\in\mathcal{S}(X)$ and $h(\pi,K_1\cup K_2)=h(\pi,K_1)\vee h(\pi,K_2)$. In other words, the index of a disjoint union is the join of the indices of the summands.

To prove Theorem 10.4 we need

Proposition 10.5

Suppose that for i=1,2, A_i is a closed set in B_i, where B_i is a topological space. Write $Y=(B_1\dot\cup B_2)/(A_1\dot\cup A_2)$, $y_0=[A_1\dot\cup A_2]$, where $\dot\cup$ denotes disjoint union with sum topology. Also, set $Z_i=B_i/A_i$, $z_i=[A_i]$, i=1,2. Then (Y,y_0) and $(Z_1,z_2)\vee(Z_2,z_2)$ are isomorphic in T^*. In other words, these spaces are homeomorphic with a base point preserving homeomorphism.

Proof of Proposition 10.5:

Let $q_i:B_i\to B_i/A_i$, i=1,2 and $q:B\to B/A$ be the quotient maps where $B=B_1\cup B_2$, $A=A_1\cup A_2$. Moreover, let

$$F_1 = B_1/A_1\times\{[A_2]\} \qquad F_2 = \{[A_1]\}\times B_2/A_2 \ .$$

Since A_i is closed in B_i, F_i is closed in $(B_1/A_1)\vee(B_2/A_2)$. If p_i is the projection of $(B_1/A_1)\vee(B_2/A_2)$ onto (B_i,A_i), define $f:(B_1/A_1)\vee(B_1/A_2)\to B/A$ as $f|F_i:=f_i\circ p_i|F_i$, $i=1,2$, where $f_i:B_i/A_i\to B/A$ is the uniquely determined continuous inclusion induced mapping (use Proposition 6.2 for that). It follows that f is well-defined, base-point preserving and continuous.

Define $G:B_1\dot{\cup}B_2\to(B_1/A_1)\vee(B_2/A_2)$ as

$$G(y_1) = (q_1(y_1),\ [A_2]) \qquad y_1\in B_1\ ,$$

$$G(y_2) = ([A_1])\ ,\ q_2(y_2)) \qquad y_2\in B_2\ .$$

G is continuous and maps A into $\{([A_1],[A_2])\}$.
Consequently there exists a unique base-point preserving continuous map $g:B/A\to(B_1/A_1)\vee(B_2/A_2)$ with $g\circ q=G$.
It is easily seen that g is the inverse of f, proving the proposition.

Proof of Theorem 10.4:

Let N_i be a strongly π-admissible isolating neighborhood of K_i, $i=1,2$. Since $K_1\cap K_2=\emptyset$, we assume that $N_1\cap N_2=\emptyset$. By Theorem 5.1 there are isolating blocks B_i, $K_i\subset B_i\subset N_i$, $i=1,2$. Then (B_i,B_i^-) is an index pair in B_i, and $h(\pi,K_i)=[B_i/B_i^-,\ [B_i^-]]$, $i=1,2$. $N:=N_1\cup N_2$ is clearly a strongly π-admissible isolating neighborhood of $K=K_1\cup K_2$, and $B:=B_1\cup B_2$ is an isolating block with $K\subset B\subset N$ and $B^-=B_1^-\cup B_2^-$. It follows that $h(\pi,K)$ is defined and $h(\pi,K)=[B/B^-,\ [B^-]]$.
By Proposition 10.5 $B/B^-=(B_1\cup B_2)/(B_1^-\cup B_2^-)$ is isomorphic (in T*) to $(B_1/B_1^-)\vee(B_2/B_2^-)$, hence in particular these two spaces have the same homotopy type. It follows that

$$h(\pi,K)=[B/B^-,\ [B^-]]=[B_1/B_1^-,\ [B_1^-]]\vee[B_2/B_2^-,\ [B_2^-]]=h(\pi,K_1)\vee h(\pi,K_2)\ .$$

The theorem is proved.

Remark:

If (Y,y_0) and (Z,z_0) are pointed spaces and $[(Y,y_0)\vee(Z,z_0)]=\overline{0}$ then $[(Y,y_0)]=[(Z,z_0)]=\overline{0}$. In fact, for a pointed space (W,x_0), $[(W,w_0)]=\overline{0}$ obviously means that (W,w_0) is contractible to a point, i.e. there is a continuous map $H:W\times[0,1]\to W$ with $H(w,0)\equiv w$, $H(w,1)\equiv w_0$ and $H(w_0,s)\equiv w_0$ for all $w\in W$, $s\in[0,1]$. If $(W,w_0)=(X,y_0)\vee(Z,z_0)$ then $W=Y\times\{z_0\}\cup\{y_0\}\times Z$, $w_0=(y_0,z_0)$ and so there is a continuous imbedding $e:(Y,y_0)\to(W,w_0)$, $e(y):=(y,z_0)$ and a continuous projection

$p:(W,w_0)\to(Y,y_0)$, $p(y,z):=y$. Define $\tilde{H}:Y\times[0,1]\to Y$ by $\tilde{H}(y,s)=pH(e(y),s)$. Then $\tilde{H}$ is a contraction of (Y,y_0) onto a point and so $[(Y,y_0)]=\overline{0}$. Similarly, $[(Z,z_0)]=\overline{0}$, and the claim follows.
This implies, in particular, that the index is "nonnegative", i.e. if the wedge sum of indices is zero, then so are the indices of the summands.
Now we consider product semiflows. Let X_1 and X_2 be two topological spaces, and π_i be a local semiflow on X_i for $i=1,2$.
Set $D=\{(t,(x_1,x_2))\in\mathbb{R}^+\times X_1\times X_2 \mid x_1\pi_1 t$ and $x_2\pi_2 t$ are defined$\}$.
Define $(x_1,x_2)\pi t=(x_1\pi_1 t,x_2\pi_2 t)$ for $(t_1,(x_1,x_2))\in D$. It is obvious that π is a local semiflow on $X_1\times X_2$ called the _product semiflow_ (or, more precisely, _product local semiflow_) of π_1 with π_2. We write $\pi=:\pi_1\times\pi_2$.
A product can analogously be defined for any finite tuple of local semiflows $(\pi_1,\ldots,\pi_p)$.
With this definition we obtain

Theorem 10.6

_If X_i is a metric space and $(\pi_i,K_i)\in\mathcal{S}(X_i)$ for $i=1,2$, then $(\pi_1\times\pi_2,\ K_1\times K_2)\in\mathcal{S}(X_1\times X_2)$ and $h(\pi_1\times\pi_2,\ K_1\times K_2)=h(\pi_1,K_1)\wedge h(\pi_2,K_2)$._

In other words, the index of a product is the smash product of the indices of the factors.

To prove Theorem 10.6 we need the following important

Theorem 10.7

For $i=1,2$. Let B_i be a topological space and $A_i\subset B_i$. Suppose that the inclusion $A_i\subset B_i$ is a cofibration for $i=1,2$. Write

$$Y = (B_1\times B_2)/(A_1\times B_2\cup B_1\times A_2),\quad y_0 = [A_1\times B_2\cup B_1\times A_2]$$

$$Z_i = B_i/A_i,\quad z_i = [A_i]\quad \text{for } i=1,2.$$

Then the pointed spaces (Y,y_0) and $(Z_1,z_1)\wedge(Z_2,z_2)$ have the same homotopy type.

Remark:

If B_i is compact for $i=1,2$ then it is easily seen (and is well-known too) that (Y,y_0) and $(Z_1,z_1)\wedge(Z_2,z_2)$ are isomorphic in T^*. This is no longer true for noncompact spaces. In fact, if these spaces were isomorphic for arbitrary topological spaces B_i and closed subspaces A_i, then this would easily imply that the smash product is associative, i.e. $Y_1\wedge(Y_2\wedge Y_3)\tilde{=}Y_1\wedge(Y_2\wedge Y_3)$. However, this latter property does not hold, in general (see Puppe [1]). On the other hand, if the pointed topological space (Y_i,y_i) have nondegenerate base points (i.e. if the inclusions $\{y_i\}\subset Y_i$ are cofibrations) then the spaces $Y_1\wedge(Y_2\wedge Y_3)$ and $(Y_1\wedge Y_2)\wedge Y_3$ are of the same homotopy type (see Puppe [1]). This latter statement is a motivation for Theorem 10.7.

Proof of Theorem 10.7:

By Exercise E6 in Chapter I of Spanier [1] the inclusion $A\subset B$, where B is a topological space and A is closed in B, is a cofibration if and only if there are continuous maps $D:B\times[0,1]\to B$ and $\Phi:B\to[0,1]$ such that:

(1) $D(x,0)=x$ for all $x\in B$

(2) $D(a,t)=a$ for all $a\in A$ and $t\in[0,1]$

(3) $\Phi^{-1}(1)=A$ and $D(\Phi^{-1}(0,1]\times\{1\})\subset A$

Let $I=[0,1]$. Assume first that $A_i\neq\emptyset$, $i=1,2$. Choose $D_i:B_i\times I\to B_i$, $\Phi_i:B_i\to I$, $i=1,2$, satisfying the above properties (1)-(3).
Let $p:B_1\times B_2\to Y$,
$q_i:B_i\to Z_i$, $i=1,2$
$q:Z_1\times Z_2\to(Z_1\times Z_2)/(\{z_1\}\times Z_2\cup Z_1\times\{z_2\})=:Z$ be the quotient maps.
For $(x_1,x_2)\in B_1\times B_2$ let

$$g(x_1,x_2) = q((q_1(x_1),\ q_2(x_2))\in Z\ .$$

Then there is a unique continuous base-point preserving mapping $\bar{g}:Y\to Z$ such that $\bar{g}\circ p=g$.
Define $f:Z_1\times Z_2\to Y$ such that for $(x_1,x_2)\in B_1\times B_2$:

$$f((q_1(x_1),\ q_2(x_2))=p(D_1(x_1,1),\ D_2(x_2,1)).$$

It is easily seen that f is well-defined.
We will now prove that f is continuous.

Since Φ_i maps A_i onto 1, there is a unique continuous map $\bar{\Phi}_i:Z_i\to I$ with $\bar{\Phi}_i\circ q_i=\Phi_i$, $i=1,2$.

Let $C_1=\{(\xi_1,\xi_2)\in Z_1\times Z_2\mid\bar{\Phi}_1(\xi_1)\geq 1/2$ or $\bar{\Phi}_2(\xi_2)\geq 1/2\}$

and $C_2=\{(\xi_1,\xi_2)\in Z_1\times Z_2\mid\bar{\Phi}_1(\xi_1)\leq 1/2$ and $\bar{\Phi}_2(\xi_2)\leq 1/2\}$.

Since $\bar{\Phi}_i$, $i=1,2$, is continuous, C_1 and C_2 are closed in $Z_1\times Z_2$.
Let $f_i=f\mid C_i$, $i=1,2$.
Obviously $f_1(\xi_1,\xi_2)\equiv y_0$, so f_1 is continuous. Since $q_1\times q_2$ maps $\{(x_1,x_2)\in B_1\times B_2\mid\Phi_1(x_1)\leq\frac{1}{2},\ \Phi_2(x_2)\leq\frac{1}{2}\}$ homeomorphically onto C_2, it follows that f_2 is continuous.
Therefore, f is continuous.

Now, there is a unique continuous base-point preserving map $\overline{f}$ with $\overline{f}oq=f$. We will show that $\overline{f}$ is a homotopy inverse of $\overline{g}$.
Let

$H:(B_1 \times B_2)\times I \to Y$

$H((x_1,x_2),t) = p(D_1(x_1,t),\ D_2(x_2,t)))$.

By 0.8 Proposition in Switzer [1] H induces a continuous map $H':Y\times I\to Y$ such that

$H'(p(x_1,x_2),t) \equiv H((x_1,x_2),t)$.

H' is, of course, base-point preserving. Furthermore, for $(x_1,x_2)\in B_1\times B_2$:

$H'(p(x_1,x_2),0) = p(x_1,x_2)$

$H'(p(x_1,x_2),1) = p(D_1(x_1,1),\ D_2(x_2,1)) = \overline{fg}(p(x_1,x_2))$.

It follows that $\overline{f}o\overline{g}\sim Id_Y$.

Now, since $D_i(a,t)\equiv a$ for $a\in A_i$, there is a unique continuous map $k_i:Z_i\times I\to Z_i$ such that $k_i(q_i(x),t)=q_i(D_i(x,t))$, i=1,2. (We use 0.8 Proposition in Switzer [1] for that).

Define $\tilde{H}:(Z_1\times Z_2)\times I\to Z$ as $\tilde{H}((\xi_1,\xi_2),t)=q(k_1(\xi_1,t),k_2(\xi_2,t))$. Then there exists a continuous map $\tilde{H}':Z\times I\to Z$ such that

$\tilde{H}'(q(\xi_1,\xi_2),t) \equiv \tilde{H}((\xi_1,\xi_2),t)$.

$\tilde{H}'$ is base-point preserving.

Moreover, if $\xi_1=q_1(x_1)$, $\xi_2=q_2(x_2)$, we have:

$\tilde{H}'(q(\xi_1,\xi_2),0) \equiv \tilde{H}((\xi_1,\xi_2),0) \equiv q(\xi_1,\xi_2)$

$$\begin{aligned}\tilde{H}'(q(\xi_1,\xi_2),1) &\equiv \tilde{H}((\xi_1,\xi_2),1) = q(q_1(D_1(x_1,1)),\ q_2(D_2(x_2,1))\\ &= \overline{g}o\overline{f}(q(\xi_1,\xi_2)).\end{aligned}$$

Hence $\overline{g}o\overline{f}\sim Id_Z$. This proves that $\overline{f}$ is a homotopy inverse of $\overline{g}$, and hence the theorem is proved for $A_i\neq\emptyset$, i=1,2.

Consider now the general case, and to this end, define spaces $\tilde{B}_i$, sets $\tilde{A}_i$, and mappings $\tilde{D}_i,\tilde{\Phi}_i$ as follows:

If $A_i\neq\emptyset$, then let $\tilde{A}_i=A_i$, $\tilde{B}_i=B_i$, $\tilde{D}_i=D_i$, $\tilde{\Phi}_i=\Phi_i$, i=1,2. If $A_i=\emptyset$, let $p_i\notin B_i$ be a point, $\tilde{A}_i=\{p_i\}$, $\tilde{B}_i=B_i\cup\{p_i\}$(with the sum topology).

$$\tilde{D}_i(x,t) \equiv x \quad \text{for } x\in\tilde{B}_i,\ t\in I$$

$$\tilde{\Phi}_i(x) = \begin{cases}0 & \text{if } x\in B_i\\ 1 & \text{if } x=p_i\end{cases} .$$

Obviously $\tilde{D}_i$, $\tilde{\Phi}_i$ satisfy conditions (1)-(3) in the remark above, so $\tilde{A}_i \subset \tilde{B}_i$ is a cofibration for i=1,2.
Now, by what has been proved so far, we get that $(\tilde{Y},\tilde{y}_0)$ and $(\tilde{Z}_1,\tilde{z}_1)\wedge(\tilde{Z}_2,\tilde{z}_2)$ are of the same homotopy type, where

$$\tilde{Y} = (\tilde{B}_1\times\tilde{B}_2)/(\tilde{A}_1\times\tilde{B}_2\cup\tilde{B}_1\times\tilde{A}_2), \quad \tilde{y}_0 = [\tilde{A}_1\times\tilde{B}_2\cup\tilde{B}_1\times\tilde{A}_2]$$

$$\tilde{Z}_i = \tilde{B}_i/\tilde{A}_i, \quad \tilde{z}_i = [\tilde{A}_i], \qquad \text{for } i=1,2.$$

But note that, by the definition of B_i/A_i, $(\tilde{Z}_i,\tilde{z}_i)=(Z_i,z_i)$. Assume now that, say, $A_1=\emptyset$. Then

$$\tilde{Y} = (B_1\times\tilde{B}_2\cup\{p_1\}\times\tilde{B}_2)/(B_1\times\tilde{A}_2\cup\{p\}\times\tilde{B}_2).$$

Thus $\tilde{Y}$ is clearly homeomorphic to $(B_1\times B_2)/(B_1\times\tilde{A}_2)$ (with base-point preserving inclusion induced homeomorphism).
This proves the theorem for $A_1=\emptyset$, $A_2\neq\emptyset$.
The same proof works for $A_2=\emptyset$, $A_1\neq\emptyset$.
If $A_1=\emptyset=A_2$, then

$$\tilde{Y} \cong (B_1\times B_2\cup B_1\times\{p_2\})/(B_1\times\{p_2\}) = (B_1\times B_2\cup\{p\})/\{p\} = (B_1\times B_2)/\emptyset = Y$$

for some $p\notin B_1\times B_2$. Hence the theorem is true, again.
The proof is complete.

<u>Proof of Theorem 10.5</u>:

For i=1,2 choose an isolating block B_i with $K_i\subset B_i\subset N_i$, where N_i is a strongly π_i-admissible isolating neighborhood of K_i. Then clearly $B=B_1\times B_2$ is an isolating block relative to $\pi_1\times\pi_2$ with $B^-=(B_1^-\times B_2)\cup(B_1\times B_2^-)$, and B is a strongly $\pi_1\times\pi_2$-admissible isolating neighborhood of $K=K_1\times K_2$.
Consequently $(\pi_1\times\pi_2, K_1\times K_2)\in S(X_1\times X_2)$ and $h(\pi_1\times\pi_2,K_1\times K_2)=[B/B^-,[B^-]]$.
However, $B_i^-\subset B_i$ is a cofibration by Theorem 3.7, i=1,2. It follows from Theorem 10.6 that

$$[B/B^-,[B^-]] = [B_1/B^-,[B_1^-]]\wedge[B_2/B_2^-,[B_2^-]] = h(\pi_1,K_2)\wedge h(\pi_2,K_2)$$

and the theorem follows.

For the rest of this section, let H_q (resp. H^q) $q\in Z$, be any unreduced homology (resp. cohomology) theory.
If $(\pi,K)\in S(X)$, $N\in N(\pi,K)$ and $\langle N_1,N_2\rangle$ is a quasi-index pair in N (rel. to π), then by the homotopy invariance property of homology (resp. cohomology), the groups $H_q(N_1/N_2,[N_2])$ (resp. $H^q(N_1/N_2,[N_2])$) do not depend on the particular choice of $N\in N(\pi,K)$ or $\langle N_1,N_2\rangle$.

Therefore we can introduce the following concept:

Definition 10.8

If $(\pi,K) \in S(X)$ then the homological index $H_q(h(\pi,K))$ (resp. the cohomological index $H^q(h(\pi,K))$ of (π,K) is defined to be $H_q(N_1/N_2,[N_2])$ (resp. $H^q(N_1/N_2,[N_2])$) for $q \in Z$ and any choice of a quasi-index pair $\langle N_1,N_2 \rangle$ in N (rel. to π), where $N \in N(\pi,K)$.
Let us now note the following useful result:

Proposition 10.9 (cf. 7.14 Proposition in Switzer [1]).

If B is a topological space and $A \subset B$ is a cofibration then

$H_q(B/A, \{[A]\}) \cong H_q(B,A)$

(resp. $H^q(B/A, \{[A]\}) \cong H^q(B,A)$), $q \in Z$.

If B is a strongly π-admissible isolating block, then $B^- \subset B$ is a cofibration by Theorem 3.7 and therefore

$$H_q(h(\pi,K)) = H_q(B,B^-) \tag{2}$$

$$\text{(resp. } H^q(h(\pi,K)) = H^q(B,B^-)) \tag{3}$$

for $q \in Z$.
Formulas (2) and (3) are a useful tool for the calculation of (co)homological indices. In particular, we will use them later on to prove a generalized Morse-equation.

1.11 Linear semiflows. Irreducibility

Let Σ^m be the homotopy type of the pointed m-dimensional sphere (S^m,s_0), $m \geq 0$. This homotopy type does not depend on the choice of the base point s_0 (see e.g. Puppe [1]). Moreover, $\Sigma^m \wedge \Sigma^n = \Sigma^{m+n}$, $m,n \geq 0$.
We have the following simple but important result:

Theorem 11.1

Let X be a Banach space and π be a global semiflow on X generating a C_0-semigroup of linear operators, i.e. assume that for every $t \geq 0$, the map $T(t):X \to X$, $T(t)x:=x\pi t$, is linear. Suppose there is a direct sum $X=X_1 \oplus X_2$, with $T(t)[X_i] \subset X_i$, $t \geq 0$, $i=1,2$, X_1 is finite dimensional, $T(t)|X_1$ can be uniquely extended to $t<0$ to form a C_0-group on X_1, and there are constants $M,\beta>0$ such that

$$\begin{aligned} \|T(t)x\| &\leq Me^{-\beta t}\|x\| \quad \text{for } x \in X_2, \quad t \geq 0, \\ \|T(t)x\| &\leq Me^{+\beta t}\|x\| \quad \text{for } x \in X_1, \quad t \leq 0 . \end{aligned} \tag{1}$$

Under these assumptions, $\{0\}$ is the largest bounded invariant set (for π), $h(\pi,\{0\})$ is defined and equal Σ^k, where $k=\dim X_1$.

Proof:

Let $P_i: X \to X$ be the projector on X_i, $i=1,2$. Let $t \to x(t)$ be a full bounded solution of π, $x_i(t) := P_i x(t)$, $i=1,2$, $t \in \mathbb{R}$.
Then $x_2(t) = x_2(t-s)\pi s$ for $t \in \mathbb{R}$, $s \in \mathbb{R}^+$.
Then (1) implies $\|x_2(t)\| \le Me^{-\beta s}\|x_2(t-s)\|$. Using the boundedness of x and letting $s \to \infty$, we obtain $x_2(t) \equiv 0$. A similar argument shows $x_1(t) \equiv 0$. This proves our first claim.
Now let N be any closed bounded set, $x^n\pi[0,t^n] \subset N$, $t^n \to \infty$ as $n \to \infty$. Write $x^n(t) := x^n\pi t$, $x_i^n(t) = P_i x^n(t)$. Let α be the Kuratowski measure of noncompactness on X (see e.g. Deimling [1]).
Since $x_1^n(t) \in P_1[N]$ for $t \in [0,t^n]$ and X_1 is finite dimensional $\{x_1^n(t^n) \mid n \ge 1\}$ is relatively compact. We thus obtain

$$\begin{aligned} &\alpha\{x^n(t^n) \mid n \ge 1\} \le \\ &\le \alpha\{x_1^n(t^n) \mid n \ge 1\} + \alpha\{x_2^n(t^n) \mid n \ge 1\} = \\ &= \alpha\{x_2^n(t^n) \mid n \ge 1\} = \alpha\{x_2^n(t^n) \mid t_n \ge s\} \\ &\le Me^{-\beta s}\alpha\{x_2^n(t^n - s) \mid t_n \ge s\} . \end{aligned}$$

Choosing $s>0$ so large that $k = Me^{-\beta s} < 1$ and applying an obvious induction argument we obtain for every $m \in \mathbb{N}$

$$\alpha\{x^n(t^n) \mid n \ge 1\} \le k^m \alpha(N) .$$

Letting $m \to \infty$ we obtain $\alpha\{x^n(t^n) \mid n \ge 1\} = 0$, i.e. $\{x^n(t^n) \mid n \ge 1\}$ has a convergent subsequence. This means that every closed bounded set is strongly π-admissible. In particular, $h(\pi,\{0\})$ is defined. To compute it, proceed as follows:
Let $\Phi: \mathbb{R}^k \to X_1$ be a linear isomorphism and let $A_1: X_1 \to X_1$ be the infinitesimal generator of $T(t)|X_1$, $t \in \mathbb{R}$. Then there is a $k \times k$-matrix C such that $\Phi^{-1}A_1\Phi = C$. From the second inequality in (1) it follows that all eigenvalues of C have positive real parts. Hence there exists a positive-definite matrix D such that $C^T D + DC = I$, where I is the identity matrix.
Now choose $\tau > 0$ such that $M \cdot (t+1)e^{-\beta t} < \frac{1}{2}$ for $t \ge \tau$. Define

$$V^+(x) = (\Phi^{-1}P_1 x)^T D(\Phi^{-1}P_1 x)$$

$$V^-(x) = \sup_{0 \le t \le \tau} \{(t+1)\|T(t)P_2 x\|\}.$$

Then V^+ and V^- are Lipschitz continuous. Moreover, an easy computation shows that

$$\liminf_{t\to 0^+} 1/t(V^+(T(t)x)-V^+(x)) > 0 \quad \text{if } x \notin X_2$$

and

$$\limsup_{t\to 0^+} 1/t(V^-(T(t)x)-V^-(x)) < 0 \quad \text{if } x \notin X_1 .$$

Define

$$B_1 = \{x \in X_1 \mid V^+(x) \leq 1\}$$

$$B_2 = \{x \in X_2 \mid V^-(x) \leq 1\}$$

$$B = B_1 \oplus B_2 .$$

By Definition 3.3 B is an isolating block for π, and $B^- = \partial B_1 \oplus B_2$. By Theorem 10.1, $h(\pi,\{0\}) = [B/B^-,[B^-]]$. Define $H: B\times[0,1]\to B$ by $H(x,\sigma) = P_1x + (1-\sigma)P_2x$. Since H maps $B^-\times[0,1]$ into B^-, H induces a continuous base-point preserving map $\tilde{H}: B/B^-\times[0,1]\to B/B^-$. $\tilde{H}$ is easily seen to be a strong deformation retraction of B/B^- onto $B_1/\partial B_1$ in the sense of Spanier [1]. Consequently $[B/B^-,[B^-]] = [B_1/\partial B_1,[\partial B_1]]$. However, the pair $(B_1,\partial B_1)$ is homeomorphic to the pair (K^k,S^{k-1}) where K^k is the k-dimensional unit ball in $\mathbb{R}^k$. Now $(K^k/S^{k-1},[S^{k-1}])$ is homeomorphic to (S^k,s_0).
This proves that $[B/B^-,[B^-]] = \Sigma^k$, completing the proof of the theorem.

Corollary 11.2

Let A be a sectorial operator on X having compact resolvent and suppose that $\sigma(A) = \sigma_1 \dot{\cup} \sigma_2$, where

$$\mathrm{re}\,\sigma_1 < -\delta < 0 \quad \text{and} \quad \mathrm{re}\,\sigma_2 > \delta > 0 \quad \text{for some } \delta > 0 .$$

Let $0 \leq \alpha < 1$ and π be the linear semiflow on X generated by solutions of the linear equation

$$\dot{u} + Au = 0 .$$

Then $K = \{0\}$ is an isolated invariant set for π, $h(\pi,\{0\})$ is defined and

$$h(\pi,\{0\}) = \Sigma^k$$

where k *is the total algebraic multiplicity of all eigenvalues* λ *of* A *with* $\mathrm{re}\sigma<0$.

Proof:

From results of Section 1.5 in Henry [1] it follows that there is a decomposition $X=X_1\oplus X_2$ and $A=A_1\oplus A_2$ with $\dim X_1=k<\infty$, $A_1:X_1\to X_1$ bounded, $\sigma(A_1)=\sigma_1$, $A_2:X_2\cap D(A)\to X_2$ sectorial on X_2 and $\sigma(A_2)=\sigma_2$. Moreover, estimates (1) as in Theorem 11.1 hold. Now Theorem 11.1 implies the corollary.

From now on, for the rest of this section, unless otherwise specified, π is an arbitrary local semiflow on an arbitrary metrix space X.

Given an invariant set K, K may contain other invariant subsets of interest, e.g. equilibria or periodic orbits. Given two such invariant subsets K_1 and K_2, the question arises whether they are connected by means of a full orbit σ in K. In other words is there a full solution $\sigma:\mathbb{R}\to K$ such that the α-limit and ω-limit sets of σ are nonempty and one of these sets is included in K_1, the other in K_2 ? An analysis of this question for simple examples leads to the following

Definition 11.3

A pair (π,K) in $\mathcal{S}=\mathcal{S}(X)$ is called *reducible*, if K is the union of two nonempty compact sets K_1 and K_2 (both these sets are then necessarily invariant) such that $h(\pi,K_1)\neq\overline{0}$ and $h(\pi,K_2)\neq\overline{0}$.
Otherwise (π,K) is called *irreducible*.
Obviously if K is connected and $(\pi,K)\in\mathcal{S}$, then (π,K) is irreducible. By the nonnegativity of the index (see Section 10) we have the following

Lemma 11.4

If $(\pi,K)\in\mathcal{S}$ *and* $h(\pi,K)=\overline{0}$, *then* (π,K) *is irreducible*.
The following result motivates the concept of irreducibility:

Theorem 11.5

Let $(\pi,K)\in\mathcal{S}$ *and assume that* (π,K) *is irreducible*. *Suppose* $K'\subset K$, *and* K' *is an isolated invariant set such that* $h(\pi,K')\neq\overline{0}$ *and* $h(\pi,K')\neq h(\pi,K)$. *Then there is a full solution* $\sigma:\mathbb{R}\to K$ of π *with* $\sigma[\mathbb{R}]\not\subset K'$ *but either* $\omega^*(\sigma)\subset K'$ *or* $\omega(\sigma)\subset K'$ (*or may be both*).

Remarks:

Theorem 11.5 says that although σ is not fully contained in K' it either emanates from K' or tends to K' (or both).
Note that Theorem 11.5 gives no means to decide which of the cases:

$\omega^*(\sigma)\subset K'$ or $\omega(\sigma)\subset K'$ actually occurs. To answer this question, additional information is needed. In fact, consider the trivial example of the local semiflow π on $X=\mathbb{R}$ generated by the solutions of the scalar ODE

$$\dot{x} = x(x-1)$$

(see Fig. 4)

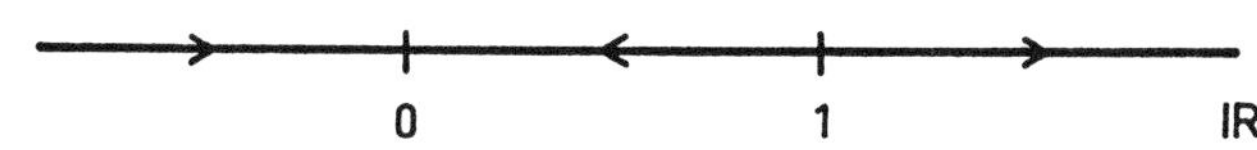

Figure 4

Let K be the interval [0,1]. Then from Fig. 4 $h(\pi,K)=\overline{0}$, $h(\pi,\{0\})=\Sigma^0$ and $h(\pi,\{1\})=\Sigma^1$. Then choosing $K'=\{0\}$ or $K'=\{1\}$ we see in both cases that the assumptions of Theorem 11.5 are satisfied. However, the unique nonconstant solution σ in K tends to K' in the first case, while it emanates from K' in the second case. Here the fact that {0} is an attractor is needed to determine the direction of the heteroclinic orbit σ.

Proof of Theorem 11.5

Let $K_1=K'$ and $K_2=K\setminus K_1$. Then K_2 is not closed. In fact, K_2 closed implies that K_2 is compact. Thus $K=K_1\dot{\cup}K_2$ and so, K_2 is isolated invariant. By irreducibility, $h(\pi,K_1)=\overline{0}$ or $h(\pi,K_2)=\overline{0}$. By our assumption, $h(\pi,K_1)\neq\overline{0}$.
Thus Theorem 10.4 implies $h(\pi,K)=h(\pi,K_1)\vee h(\pi,K_2)=h(\pi,K)\vee\overline{0}=h(\pi,K_1)$, a contradiction.
Thus, indeed, K_2 is not closed. Consequently, there is a sequence $x_n\in K_2$ converging to some $x_0\in K_1$. Let N be a strongly π-admissible isolating neighborhood of K_1. Moreover, for every $n\in\mathbb{N}$, let $\sigma_n:\mathbb{R}\to K$ be a full solution through x_n. We may assume that $x_n\in \operatorname{Int} N$ for all $n\geq 1$. Since K_1 is maximal invariant in N and $x_n\notin K_1$ for $n\geq 1$, it follows that $\sigma_n[\mathbb{R}]\not\subset N$. Taking subsequences if necessary we have the following two cases:

1. Case:

For every $n\geq 1$ there is a $t_n>0$ such that

$$\sigma_n[0,t_n]\subset N\cap K, \quad \sigma_n(t_n)\in\partial N.$$

Since $x_0 \in K_1$ and K_1 is invariant it follows that $t_n \to \infty$. Therefore we may assume that $\sigma_n(t_n)$ converges to some $y_0 \in A^-(K \cap N) \cap \partial N \cap K$, i.e. there is a solution $\sigma: \mathbb{R} \to K$ through y_0 such that $\sigma(\mathbb{R}^-) \subset N \cap K$. Since the α-limit set of σ is invariant it follows that $\omega^*(\sigma) \subset K_1$ and $\sigma(\mathbb{R}) \not\subset K_1$ (since $\sigma(0) \notin K_1$).

2. Case:

For every $n \geq 1$ there is a $t_n < 0$ such that

$$\sigma_n[t_n, 0] \subset N \cap K \quad \text{and} \quad \sigma_n(t_n) \in \partial N .$$

Since K is compact, w.l.o.g. $y_n = \sigma_n(t_n)$ converges to some $y_0 \in N \cap K \cap \partial N$. Moreover, $y_n \pi [0, t_n] \subset N \cap K$ for $n \geq 1$. If $\{t_n\}$ is bounded, then w.l.o.g. $t_n \to t_0$ as $n \to \infty$, so $y_n \pi t_n = x_n \to y_0 \pi t_0$. Hence $x_0 = y_0 \pi t_0$, and thus $y_0 \in A^+(N \cap K) \cap \partial N$.
If $\{t_n\}$ is unbounded, then w.l.o.g. $t_n \to \infty$ and so again $y_0 \in A^+(N \cap K) \cap \partial N$. Let $\sigma: \mathbb{R} \to K$ be a full solution through y_0. It follows that $\omega(\sigma) \subset K_1$ and $\sigma[\mathbb{R}] \not\subset K_1$.
The theorem is proved.

Theorem 11.6

If $(\pi, K) \in S$ and $h(\pi, K) = \Sigma^m$ for some $m \geq 0$, then (π, K) is irreducible.
We need

Lemma 11.7

Let (Y, y_0) and (Z, z_0) be two pointed spaces and $(W, w_0) := (Y, y_0) \vee (Z, z_0)$. If $h(W, w_0) = \Sigma^m$ for some $m \geq 0$, then either $[(Y, y_0)] = \bar{0}$ or $[(Z, z_0)] = \bar{0}$.

Proof of the lemma:

First assume $m \geq 1$ and let $\pi_n(A, a_0)$ be the n-th homotopy group of (A, a_0), $n \geq 1$. Since $\pi_n(Y, y_0) \oplus \pi_n(Z, z_0)$ imbeds injectively into $\pi_n(W, w_0)$ (see Exercise B.6, Chapter 7 in Spanier [1]), and since $\pi_m(S^m, s_0) = Z$, it follows that $\pi_m(Y, y_0) = 0$ or $\pi_m(Z, z_0) = 0$. Assume w.l.o.g. that $\pi_m(Y, y_0) = 0$. Since (W, w_0) and (S^m, s_0) have the same homotopy type, there are maps $f: (Y, y_0) \to (S^m, s_0)$ and $g: (S^m, s_0) \to (Y, y_0)$ such that $g \circ f$ is homotoppic to $Id_{(Y, y_0)}$. Since $\pi_m(Y, y_0) = 0$, g is null-homotopic, so $g \circ f$ is null-homotopic, which implies that $Id_{(Y, y_0)}$ is null-homotopic i.e. (Y, y_0) is contractible. This proves the lemma for $m \geq 1$.
If m=0, the lemma follows by considering the maps

$$f: (W, w_0) \to (\{-1, +1\}, -1)$$
$$g: (\{-1, +1\}, -1) \to (W, w_0)$$

such that $g \circ f$ and $f \circ g$ are homotopic to the corresponding identity maps. Hence, in fact, $f \circ g$ is the identity map. Consequently g is injective and w.l.o.g. $g(1)=(y',z_0)$, $y' \neq y_0$. Using the homotopy from $g \circ f$ to $Id_{(W,w_0)}$ it is now easily seen that (Z,z_0) is contractible. The lemma is proved.

Proof of Theorem 11.6:

If the theorem is not true, then there is a disjoint union $K=K_1 \dot{\cup} K_2$ with $h(\pi,K_1)\neq\overline{0}$, $h(\pi,K_2)\neq\overline{0}$. But $h(\pi,K)=h(\pi,K_1)\vee h(\pi,K_2)$ by Theorem 10.3 Lemma 11.7 implies that $h(\pi,K_1)=\overline{0}$ or $h(\pi,K_2)=\overline{0}$, a contradiction. We will end this section by a result which gives conditions assuring that $A^-(N)\neq A(N)=K$.

Theorem 11.8

Let $(\pi,K)\in \mathcal{S}(X), K\neq\emptyset$, and let $h(\pi,K)$ be the homotopy type of a connected pointed space (Y,y_0). Then for every isolating neighborhood N of K it follows that $K=A(N)\neq A^-(N)$.

Proof:

Suppose that $K=A^-(N)$, for some isolating neighborhood N of K. Then $K=A^-(\tilde{N})$, for every isolating neighborhood $\tilde{N}\subset N$ of K. Therefore we can assume that N is strongly π-admissible.
By Corollary 5.5, there is an isolating block B, $K\subset B\subset N$, with $B^-=\emptyset$. Then $(B/B^-,[B^-])\cong(B\dot{\cup}\{p\},p\}$ where $p\notin B$. Then, by Theorem 10.1 $h(\pi,K)=[(B\dot{\cup}\{p\},p)]=[(Y,y_0)]$. Let $\overline{H}^q$ be Alexander-Spanier cohomology theory with coefficients in Z. Then by Theorem 5 in Section 4 of Chapter 6 in Spanier [1], for every pointed set (W,w_0), $\overline{H}^0(W,\{w_0\})$ is isomorphic to the module of all locally constant functions f from W to Z with $f(w_0)=0$.
Now, from our assumption

$$\overline{H}^q(B\dot{\cup}\{p\},\{p\}) \cong \overline{H}^q(Y,\{y_0\}), \quad q\in Z.$$

Since Y is connected, every locally constant function on Y is constant. Consequently, $\overline{H}^0(Y,\{y_0\})=\{0\}$. However, since $B\neq\emptyset$, $\overline{H}^0(B\cup\{p\},\{p\})$ contains, e.g. the locally constant function $f:B\dot{\cup}\{p\}\to Z$ defined as $f(x)=1$ if $x\in B$, $f(p)=0$. Therefore, $\overline{H}^0(B\dot{\cup}\{p\},\{p\})\neq\{0\}$, a contradiction.

We now obtain the following corollary:

Corollary 11.9

Let π be a global semiflow on a normed linear space X. Let K_∞ be the union of all full bounded orbits of π, i.e. $K_\infty=\{x \in X \mid$ there is a full solution σ of π through x with $\sigma[\mathbb{R}]$ bounded$\}$.
Assume that $K_\infty \neq \emptyset$ and $(\pi, K_\infty) \in S(X)$. Suppose also that $h(\pi, K_\infty)$ is the homotopy type of a pointed space (Y, y_0) with Y connected. Then there exists a full solution $\sigma:\mathbb{R}\to X$ of π with $\sup_{t\leq 0}\|\sigma(t)\|<\infty$ and $\sup_{t\geq 0}\|\sigma(t)\|=\infty$.

Proof:

Let N be a strongly π-admissible isolating neighborhood of K_∞. Suppose the corollary is not true. Let $x \in A^-(N)$. Then there is a solution $\tilde{\sigma}:\mathbb{R}^- \to N$ of π with $\tilde{\sigma}(0)=x$. Since π is global, we have $\omega_x=\infty$ and so σ can be extended to a full solution through x. Since $A^-(N)$ is compact by Theorem 4.5 we obtain that $\sigma[\mathbb{R}^-]$ is bounded. Therefore $\sigma[\mathbb{R}^+]$ is also bounded, and so $x \in K_\infty$. It follows that $A^-(N)=A(N)=K$. Now Theorem 11.8 immediately implies a contradiction, proving the corollary.

1.12 Continuation of the homotopy index

There is a formal similarity between the homotopy index and the Leray-Schauder or, more generally, the Nussbaum fixed point index. In fact let X be some Banach space and Ω be open and bounded in X, and $f:\mathrm{Cl}\Omega\to X$ be a compact or more generally, a locally α-condensing map. If there are no fixed points of f on $\partial\Omega$, i.e. if $f(x)\neq x$ for $x \in\partial\Omega$, then the fixed point index $i(f,\Omega)$ is defined and measures "algebraically" the set of all fixed points in Ω. Similarly if π is a local semiflow on a neighborhood of $\mathrm{Cl}\Omega$ in X and $\mathrm{Cl}\Omega$ is π-admissible (which is an asymptotic compactness condition like α-condensing and α-contracting, cf. Rybakowski [4]), then $h(\pi,K)$ is defined where $K=A_\pi(\mathrm{Cl}\Omega)$. Here, $h(\pi,K)$ "measures" the largest invariant set in $\mathrm{Cl}\Omega$.
If $i(f,\Omega)\neq 0$, then there is at least one fixed point of f in Ω. Similarly, if $h(\pi,K)\neq\overline{0}$, then $K\neq\emptyset$. The fixed point index is homotopy invariant: if f depends continuously on a parameter $\lambda \in [0,1]$, f is "collectively" an α - contraction and $f(\lambda,x)\neq x$ for all $x \in \partial\Omega$ $\lambda \in [0,1]$, then $i(f_\lambda,\Omega)$ does not depend on λ (cf. Deimling [1]). This homotopy invariance of the fixed point index is its most important property: It makes this index a crucial tool in perturbation problems.
Now, the homotopy index also has the same property. Roughly speaking,

the homotopy invariance of the index can be described as follows: Suppose that the semiflows π_λ on X vary "continuously" for $\lambda \in [0,1]$. If N is some closed set, if $K_\lambda := A_{\pi_\lambda}(N) \subset \mathrm{Int}\, N$ (i.e. if N isolates K_λ for every λ), and if certain admissibility conditions are satisfied, then $h(\pi_\lambda, K_\lambda) \equiv$const. for all $\lambda \in [0,1]$.
The homotopy invariance of the homotopy index is proved in this section. In Chapter III we show how the categorial Morse index is "continued". Although both continuation properties follow from the same general results, it is instructive to give the proofs first for the easier case of the homotopy index.
First we need a precise description of what it means that a pair (π_λ, K_λ) varies "continuously" as λ varies in Λ, Λ being [0,1] or more generally, some metric space. Our definition will agree with the intuitive description given above.
For the rest of this section, X is a fixed metric space. Recall that by $S=S(X)$ we denote the following class: $S=S(X)=\{(\pi,K) \mid \pi$ is a local semiflow on X, K is a closed π-invariant set and there is a closed set N such that $K=A_\pi(N) \subset \mathrm{Int}\, N$ and N is strongly π-admissible$\}$.
As we know from earlier sections, S is precisely the class of pairs (π,K) for which the categorial Morse index and the homotopy index are defined.

<u>Definition 12.1</u>

Let $\alpha: \Lambda \to S$ be a mapping from a metric space Λ into $S=S(X)$. We will write $(\pi_\lambda, K_\lambda) := \alpha(\lambda)$ for $\lambda \in \Lambda$. α is called S-<u>continuous</u>, if for every $\lambda_0 \in \Lambda$ there is a neighborhood W of λ_0 in Λ and a closed set N in X such that:

(1) For every $\lambda \in W$, N is a strongly π_λ-admissible isolating neighborhood of K_λ, relative to π_λ.
(2) Whenever $\lambda_n \to \lambda_0$, then $\pi_{\lambda_n} \to \pi_{\lambda_0}$ as $n \to \infty$ and N is $\{\pi_{\lambda_n}\}$-admissible.

The main result of this section is

<u>Theorem 12.2</u>

<u>If</u> $\alpha: \Lambda \to S$ <u>is</u> S-<u>continuous then</u> $h(\pi_\lambda, K_\lambda)$ <u>is constant for</u> λ <u>lying in connected components of</u> Λ. <u>In particular, if</u> $\Lambda = [0,1]$, $h(\pi_\lambda, K_\lambda) \equiv$<u>const. for all</u> $\lambda \in \Lambda$.

The rest of this section will be devoted to the proof of this result: in fact, Theorem 12.2 is a consequence of the following

Theorem 12.3

Suppose π_k, $k \in \mathbb{N}$ and π_0 are local semiflows on X. Suppose $\tilde{N}$ is a closed set in X, and for every $k \in \mathbb{N} \cup \{0\}$, $\tilde{N}$ is strongly π_k-admissible. Moreover, assume that $\pi_n \to \pi_0$ as $n \to \infty$ and that for every subsequence $\{\pi_{n_m}\}$ of $\{\pi_n\}$, $\tilde{N}$ is $\{\pi_{n_m}\}$-admissible.

Let $K_n = A_{\pi_n}(\tilde{N})$, $n \in \mathbb{N} \cup \{0\}$, and assume $K_0 \subset \mathrm{Int}\,\tilde{N}$. Then there exist two closed sets N' and N and $n_0 \in \mathbb{N}$ and for each $n \in \{n \geq n_0\} \cup \{0\}$, there exist two pairs of sets $\langle N'_{1,n}, N'_{2,n}\rangle$ and $\langle N_{1,n}, N_{2,n}\rangle$ such that the following properties hold for n=0 and $n \geq n_0$:

(1) $K_n \subset N' \subset \mathrm{Int}\,N \subset \mathrm{Int}\,\tilde{N}$, $N'_{1,0} = N'$, $N_{1,0} = N$.

(2) $\langle N'_{1,n}, N'_{2,n}\rangle$ *(resp.* $\langle N_{1,n}, N_{2,n}\rangle$*) is an index pair in* N' *(resp. in* N*) relative to* π_n.

(3) $N'_{1,n} \subset N'_{1,0} \subset N_{1,n} \subset N_{1,0}$,

$N'_{2,n} \subset N'_{2,0} \subset N_{2,n} \subset N_{2,0}$.

Proof of Theorem 12.2:

Let $\lambda_0 \in \Lambda$ be arbitrary. By Definition 12.1, there is a neighborhood $\tilde{W}$ of λ_0 in Λ and a closed set $\tilde{N}$ such that (1) and (2) of that definition hold with N replaced by $\tilde{N}$. We will prove that there is an open set $W \subset \tilde{W}$, $\lambda_0 \in W$, such that $h(\pi_\lambda, K_\lambda) \equiv \mathrm{const} = h(\pi_{\lambda_0}, K_{\lambda_0})$ for all $\lambda_0 \in W$. In fact, if this is not true, then there is a sequence $\lambda_n \to \lambda_0$, $\lambda_n \in \tilde{W}$ such that $h(\pi_n, K_n) \neq h(\pi_0, K_0)$ where $\pi_n = \pi_{\lambda_n}$ $K_n = K_{\lambda_n}$, $n \in \mathbb{N} \cup \{0\}$. Then all assumptions of Theorem 12.3 are satisfied. Therefore, by Theorem 12.3 and results of Section 1.9 we have for every $n \geq n_0$ a sequence of continuous inclusion induced mappings

$$N'_{1,n}/N'_{2,n} \xrightarrow{i_1} N'_{1,0}/N'_{2,0} \xrightarrow{i_2} N_{1,n}/N_{2,n} \xrightarrow{i_3} N_{1,0}/N_{2,0} .$$

Thus $i_2 \circ i_1$ and $i_3 \circ i_2$ are inclusion induced and by Definition 9.2 $i_2 \circ i_1 \in \mathcal{V}(\pi_n, K_n)$, $i_3 \circ i_2 \in \mathcal{V}(\pi_0, K_0)$.

By Theorem 9.4 $[i_2 \circ i_1]$ and $[i_3 \circ i_2]$ are isomorphisms in the homotopy category of pointed spaces. By the simple Lemma 12.4 below and the fact that $[i_3 \circ i_2] = [i_3] \circ [i_2]$ and $[i_2 \circ i_1] = [i_2] \circ [i_1]$, we immediately get that all maps i_1, i_2, i_3 are homotopy equivalences between pointed spaces. Hence in particular

$$[N_{1,n}/N_{2,n}, [N_{2,n}]] = [N_{1,0}/N_{1,0}, [N_{2,0}]].$$

But the left hand side of this equation is $h(\pi_n,K_n)$, and the right hand side is $h(\pi_0,K_0)$. Thus $h(\pi_n,K_n)=h(\pi_0,K_0)$, a contradiction which proves the theorem.

Lemma 12.4

Let C be any category, and a,b,c be morphisms in C. If $c\circ b$ and $b\circ a$ are defined and isomorphisms, then a,b and c are all isomorphisms.

Proof:

Let $a_1=(b\circ a)^{-1}\circ b$. Then a_1 is defined and $a_1\circ a=(b\circ a)^{-1}\circ(b\circ a)=id$. Now $(b\circ a)\circ a_1=b$. This implies $c\circ b\circ a\circ a_1=c\circ b$. It follows that $a\circ a_1=(c\circ b)^{-1}\circ(c\circ b)=id$. Thus a is an isomorphism with $a^{-1}=a_1$. But this immediately implies that b and c are isomorphisms as well. To prove Theorem 12.3, we shall need some lemmas:

Lemma 12.5

Suppose the assumptions of Theorem 12.3 are satisfied. Moreover, assume that V,U and $\tilde{U}$ are open sets such that $\tilde{N}=Cl\,\tilde{U}$, $ClV\subset U, N:=ClU\subset\tilde{U}$. Let $n\in\mathbb{N}\cup\{0\}$ be arbitrary. Define the map $t_n^+(x)=\sup\{t\geq 0 \mid x\pi_n[0,t]$ is defined and $\subset\tilde{U}\}$, for $x\in\tilde{U}$. Finally, let $K_n\subset V$. Then for every $M>0$, the sets $N_{1,n}=N\cap Cl\{y \mid$ there are $x\in V$, $t\geq 0$, such that $x\pi_n[0,t]\subset\tilde{U}$ and $x\pi_n t=y\}$, $N_{2,n}=N_{1,n}\cap\{y\in\tilde{U}\mid t_n^+(y)\leq M\}$, enjoy the following properties:

(1) *$N_{1,n}$ and $N_{2,n}$ are closed and N-positively invariant subsets of N (relative to π_n),*

(2) $K_n\subset V\cap\{y\in\tilde{U}\mid t_n^+(y)>M\}\subset N_{1,n}\setminus N_{2,n}$,

(3) $V\subset N_{1,n}$.

Proof:

If $x\in V$, then let $y=x$, $t=0$ in the definition of $N_{1,n}$, proving that (3) is satisfied.

Since $t_n^+(x)=\infty$ for all $x\in K_n$, (2) follows immediately. Obviously $N_{1,n}$ is closed. Let $x_k\in N_{2,n}$, $x_k\to x_0$ as $k\to\infty$. Then $x_0\in N_{1,n}$. Since t_n^+ is lower-semicontinuous (see Proposition 5.2), it follows that $t_n^+(x_0)\leq M$ and so $x_0\in N_{2,n}$ proving that $N_{2,n}$ is closed. Suppose now that $x\in N_{1,n}$ and s is such that $x\pi_n[0,s]$ is defined and $\subset N$. Then there are $y_k\to x$, $x_k\in V$, $t_k\geq 0$ such that $x_k\pi_n[0,t_k]\subset\tilde{U}$ and $x_k\pi_n t_k=y_k$ for all k. Since $N\subset\tilde{U}$ it follows that for k large enough, $y_k\pi_n[0,s]\subset\tilde{U}$, i.e. $x_k\pi_n[0,t_k+s]\subset\tilde{U}$. This proves that $x\pi_n[0,s]\subset N_{1,n}$. Hence $N_{1,n}$ is N-positively invariant (relative to π_n). Since t_n^+ does not increase along solutions of π_n, the same is true for $N_{2,n}$. The lemma is proved.

Lemma 12.6

Assume all hypotheses of Theorem 12.3. Moreover, assume that $K_0 \neq \emptyset$, $\tilde{N}$ *is an isolating block for* π_0, *and there is an open set* $\tilde{U}$ *with* $K_0 \subset \tilde{U}$ *and* $\tilde{N} = \mathrm{Cl}\tilde{U}$. *Define* g^- *as in Proposition 5.2 with* N (*resp.* U) *replaced by* $\tilde{N}$ (*resp.* $\tilde{U}$) *and* π *replaced by* π_0 *in that proposition. Finally, let* t_n^+, *for* $n \in \mathbb{N} \cup \{0\}$ *be as in Lemma 12.5. Then whenever* $x_n \to x_0$ *in* $\tilde{U}$, *then* $t_n^+(x_n) \to t^+(x_0)$. *Now define for* $a,b>0$,

$$V(a,b) = \{x \in \tilde{U} \mid g^-(x) < a,\ t_0^+(x) > b\}.$$

Then for some $a_0, b_0 > 0$, $N := \mathrm{Cl}V(a_0,b_0) \subset \tilde{U}$. *Let* $U = V(a_0,b_0)$. *Then for every* $M > b_0$ *there is an* $\varepsilon_0 = \varepsilon_0(M)$, $0 < \varepsilon_0 < a_0$, *such that for every* $0 < \varepsilon \leq \varepsilon_0$ *there is an* $n_0 = n_0(\varepsilon, M)$ *such that for all* $n \geq n_0$ *properties* (1) *and* (2) *below hold:*

(1) $\mathrm{Cl}V(\varepsilon,M) \subset U$, $K_n \subset V(\varepsilon,M)$,

(2) *If the sets* $N_{1,n} = N_{1,n}(\varepsilon,M)$, $N_{2,n} = N_{2,n}(\varepsilon,M)$ *are defined as in Lemma 12.5 with* V *replaced by* $V(\varepsilon,M)$, *then* $\langle N_{1,n}, N_{2,n} \rangle$ *is an index pair in* N (*relative* to π_n).

Proof:

Let $x_n \to x_0$ in $\tilde{U}$. Let M be such that $t_0^+(x_0) < M$. Then $0 < t_0^+(x_0) < \infty$, so $x_0\pi_0[0, t_0^+(x_0)) \subset \tilde{U}$, $x_0\pi_0 t_0^+(x_0) \in \partial\tilde{N}$. Since $\tilde{N}$ is an isolating block for π_0, it follows that $y_0 = x_0\pi_0 t_0^+(x_0)$ is a strict egress point and for some small $\varepsilon > 0$, $y_0\pi_0(0,\varepsilon) \subset X \setminus \tilde{N}$. Therefore for some $s < M$, $x_0\pi_0 s \notin \tilde{N}$. It follows that for n large enough, $x_n\pi_n s_0 \notin \tilde{N}$, so $s > t_n^+(x_n)$. Consequently $t_n^+(x_n) < M$.

Now let $t_0^+(x_0) > M$. Then $x_0\pi_0[0,M] \subset \tilde{U}$, and so $x_n\pi_n[0,M] \subset \tilde{U}$ for all n large, proving that $t_n^+(x_n) > M$. Thus $t_0^+(x_n) \to t_0^+(x_0)$ as claimed.

Now we define $V = V(a_0,b_0)$ as in the statement of the Lemma. V is open by Proposition 5.2. Whenever $g^-(x_n) \to 0$ there is a subsequence $\{x_{n_m}\}$ with $x_{n_m} \to x_0 \in A^-_{\pi_0}(\tilde{N})$. Hence, if also $t_0^+(x_n) \to \infty$ as $n \to \infty$, it follows that $x_0 \in K_0 \subset \tilde{U}$. This easily implies (cf. also the proof of Lemma 5.4) that $\mathrm{Cl}V(a_0,b_0) \subset \tilde{U}$ for some a_0, b_0.

Now let $M > b_0$, $\varepsilon < a_0$. Since $\tilde{N}$ is an isolating block for π_0, g^- and t_0^+ are continuous on $\tilde{U}$. Hence, obviously, $\mathrm{Cl}V(\varepsilon,M) \subset V(a_0,b_0) = U$. Moreover, $K_0 \subset V(\varepsilon,M)$.

By Theorem 4.5, part 2.2, there is an $n_0 = n_0(\varepsilon,M)$ such that for $n \geq n_0$, $K_n \subset V(\varepsilon,M)$, i.e. (1) of Lemma 12.6 is satisfied.

Therefore, by Lemma 12.5 the sets N, $N_{1,n}(\varepsilon,M)$ and $N_{2,n}(\varepsilon,M)$ satisfy the conditions of Definition 7.1 of index pairs, except may be for condition (3) in that definition. Hence, if the lemma is not true, then there is an $M_0>b_0$ and sequences
$\varepsilon_m\to\infty$, $n_m\to\infty$, $y_m\in N_{1,n_m}(\varepsilon_m,M_0)\cap\partial N$, $y_m\notin N_{2,n_m}$.
By the definition of $N_{1,n_m}(\varepsilon_m,M_0)$ for every m there are $\tilde{y}_m\in X$, $x_m\in V(\varepsilon_m,M)$ and $t_m\geq 0$ that $d(y_m,\tilde{y}_n)<2^{-m}$, $x_m\pi_{n_m}[0,t_m]\subset\tilde{U}$ and $\tilde{y}_m=x_m\pi_{n_m}t_m$. Since $g^-(x_m)\to 0$ we can assume w.l.o.g. that $x_m\to x_0\in A^-_{\pi_0}(\tilde{N})$. Now it follows from admissibility and the fact that $\pi_{n_m}\to\pi_0$ as $n\to\infty$, that $\{x_m\pi_{n_m}t_m\}$ has a convergent subsequence, w.l.o.g. $x_m\pi_{n_m}t_m\to y_0\in A^-_{\pi_0}(\tilde{N})$. It follows that $y_0\in A^-_{\pi_0}(\tilde{N})\cap\partial N$ and $y_m\to y_0$ as $m\to\infty$. As $t^+_{n_m}(y_m)>M_0$ and $\tilde{N}$ is an isolating block, it follows that $t^+_0(y_0)\geq M_0>b_0$. Since $g^-(y_0)=0$, $y_0\in U\cap\partial N=\emptyset$, a contradiction which proves the lemma.

Lemma 12.7

Assume all hypotheses of Lemma 12.6. Let $\tilde{U}$ be as in that lemma. Then for every $M>0$ there is an $\varepsilon_1=\varepsilon_1(M)$ and $n_1=n_1(M)$ such that for all $0<\varepsilon\leq\varepsilon_0$, $n\geq n_0$:

$$C_n(\varepsilon)\subset D(\varepsilon)\subset\hat{C}_n(\varepsilon)\subset\hat{D}(\varepsilon)\quad,$$

where

$$C_n(\varepsilon) = \{x\in\tilde{U}\ |\ t^+_n(x)\leq 2M\}\cap \mathrm{Cl}V(\varepsilon,M)\},$$

$$D(\varepsilon) = \{x\in\tilde{U}\ |\ t^+_0(x)\leq 3M\}\cap \mathrm{Cl}V(\varepsilon,M)\},$$

$$\hat{C}_n(\varepsilon) = \{x\in\tilde{U}\ |\ t^+_n(x)\leq 4M\}\cap N_{1,n}(\varepsilon,M)\},$$

$$\hat{D}(\varepsilon) = \{x\in\tilde{U}\ |\ t^+_0(x)\leq 5M\}\cap N\}.$$

Proof:

If the lemma is not true then, taking subsequences as necessary, we can assume that there are sequences $x_n\in X$ and $\varepsilon_n\to 0$ such that either $x_n\in C_n(\varepsilon_n)\setminus D(\varepsilon_n)$ for all n, or $x_n\in D(\varepsilon_n)\setminus\hat{C}_n(\varepsilon_n)$ for all n, or $x_n\in\hat{C}_n(\varepsilon_n)\setminus\hat{D}(\varepsilon_n)$ for all n. Using the arguments from the proof of Lemma 12.6 we see that $\{x_n\}$ has a convergent subsequence, w.l.o.g. $x_n\to x_0\in N\subset\tilde{U}$. Then $t^+_n(x_n)\to t^+_0(x_0)$ by Lemma 12.6. Now this immediately gives a contradiction, proving the lemma.

We will now prove Theorem 12.3:

<u>1. Case</u>: $K_0=\emptyset$.

Then $K_n=\emptyset$ for all n large enough:
In fact, if this is not true then w.l.o.g. $K_n\neq\emptyset$ for all $n\in\mathbb{N}$. Choose $x_n\in K_n$. Then there are $y_n\in K_n$, $t_n\to\infty$, with $y_n\pi_n t_n=x_n$. By admissibility, we can assume that $x_n\to x_0\in A^-_{\pi_0}(\tilde{N})$. (cf. Theorem 4.5).
Furthermore, $x_n\pi_n[0,t_n]\subset\tilde{N}$, so $x_0\in A^+_{\pi_0}(\tilde{N})$, i.e. $K_0=A^-_{\pi_0}(\tilde{N})\cap A^+_{\pi_0}(\tilde{N})\neq\emptyset$, a contradiction.
Choose $N=N'=\emptyset$ and $N_{1,n}=N_{2,n}=N'_{1,n}=N'_{2,n}=\emptyset$ for all n large enough and for n=0. Then properties (1)-(3) hold in this case.

<u>2. Case</u>: $K_0\neq\emptyset$.

We can assume w.l.o.g. that $\tilde{N}$ is an isolating block for K_0 relative to π_0 (otherwise apply Theorem 5.1 to get a block B, $K_0\subset B\subset\tilde{N}$). Let $\tilde{U}=\operatorname{Int}\tilde{N}$. Then $Cl\tilde{U}\subset\tilde{N}$ and $Cl\tilde{U}$ is easily seen to be an isolating block, too. Hence we can assume that there is an open set $\tilde{U}$ such that $Cl\tilde{U}=\tilde{N}$. Choose $a_0,b_0>0$ such that $ClV(a_0,b_0)\subset\tilde{U}$ (see Lemma 12.6). Let $U=V(a_0,b_0)$ $N=ClU$.
Let $M>b_0$. Choose $\varepsilon_1=\varepsilon_1(M)$ and $n_1=n_1(M)$ such that the conclusions of Lemma 12.7 hold for $\varepsilon\leq\varepsilon_1$ and $n\geq n_1$. Choose $\varepsilon_0\leq\varepsilon_1$ such that the conclusions of Lemma 12.6 hold for $\varepsilon\leq\varepsilon_0$ and $n\geq n_0(\varepsilon,M)$. Choose $\varepsilon=\varepsilon_0$ and $n\geq\max(n_1(M),\ n_0(\varepsilon_0,M))$. Then by Lemma 12.6, $N_{1,n}(\varepsilon,M)$ and $N_{2,n}(\varepsilon,M)$ form an index pair in N, relative to π_n, and $K_n\subset V(\varepsilon,M)$.

Let $\tilde{N}_{2,n}=N_{1,n}(\varepsilon,M)\cap\{x\in\tilde{U}\mid t^+_n(x)\leq 4M\}$. Since $\tilde{N}_{2,n}\supset N_{2,n}(\varepsilon,M)$, Lemmas 12.5 and 12.6 imply that $\langle N_{1,n}(\varepsilon,M),\ \tilde{N}_{2,n}\rangle$ are an index pair in N relative to π_n. Write $N_{1,n}:=N_{1,n}(\varepsilon,M)$, $N_{2,n}:=\tilde{N}_{2,n}$. Now let $U'=V(\varepsilon,M)$, $N'=ClV(\varepsilon,M)$. Applying Lemma 12.6 with U,N,a_0,b_0,M being replaced , respectively, by $U',N',\varepsilon,M,2M$, we obtain an $\varepsilon'_0=\varepsilon'_0(2M)<\min(\varepsilon_0,\varepsilon_1)$ and $n'_0=n'_0(\varepsilon'_0,2M)\geq\max(n_1(M),\ n_0(\varepsilon_0,M))$ such that for $n\geq n'_0$, the sets $N_{1,n}(\varepsilon'_0,2M)$, $N_{2,n}(\varepsilon'_0,2M)$ form an index pair in N' relative to π_n, and $K_n\subset V(\varepsilon'_0,2M)$. Write $N'_{1,n}=N_{1,n}(\varepsilon'_0,2M)$, $N'_{2,n}=N_{2,n}(\varepsilon'_0,2M)$. Then for $n\geq n'_0$:

(1) $K_n\subset N'\subset U\subset\operatorname{Int}N\subset N\subset\tilde{U}\subset\operatorname{Int}\tilde{N}$

(2) of the statement of Theorem 12.3 holds.

(3) $N'_{1,n}\subset N'\subset N_{1,n}\subset N$, $N'_{2,n}\subset\{x\in\tilde{U}\mid t^+_n(x)\leq 2M\}\cap ClV(\varepsilon,M)\subset$
$\subset\{x\in\tilde{U}\mid t^+_0(x)\leq 3M\}\cap ClV(\varepsilon,M)=:N'_{2,0}\subset N_{2,n}\subset\{x\in\tilde{U}\mid t^+_0(x)\leq 5M\}\cap N=:N_{2,0}$.

To complete the proof of Theorem 12.3, it is therefore only necessary to show that $\langle N', N'_{2,0}\rangle$ and $\langle N, N_{2,0}\rangle$ are index pairs in N' (resp. N) relative to π_0. However, this is obvious from the fact that g^- and t_0^+ are continuous and from properties proved in Proposition 5.2. The theorem is proved.

Chapter II

Applications to partial differential equations

We will now give a few applications of the theory developed in Chapter I. First we describe some of the types of differential operators which generate sectorial operators.
Then we discuss local center manifolds for parabolic equations and state an existence and approximation result for center manifolds.

Next, in Section 2.3 we prove an index product formula relating the index of a "small" invariant set relative to the PDE to the index of the same invariant set with respect to the restricted flow on a local center manifold. This formula together with the approximation result for center manifolds permits to calculate the index in critical cases. An example of this is given in Section 2.4.
In Section 2.5 we consider asymptotically linear parabolic equations with nonresonance at infinity. More generally, we roughly assume that the graph of the nonlinearity f lies, for all large values of the argument, between two consecutive eigenvalue branches of the operator A. For such equations we prove that the union of all full bounded solutions of the differential equation is compact and we compute its homotopy index. Since our parabolic equation is gradient-like, this will give us a first existence result of solutions of the corresponding elliptic boundary value problem.
If 0 is an isolated equilibrium of our equation then we can compute its index and thus give conditions assuring the existence of nontrivial solutions of the elliptic equation as well as the existence of heteroclinic orbits for the parabolic equation. This is done in Section 2.6.
Using the homotopy index and the maximum principle we then prove (in Section 2.7) the existence of positive solutions of second order elliptic boundary value problems and heteroclinic orbits of the corresponding parabolic boundary value problems.
In Section 2.8 we consider certain variational operator equations of the form $Au=f(u)$ where A is linear and noninvertible.

By combining the Liapunov-Schmidt reduction procedure with the homotopy invariance of the homotopy index we arrive at a continuation principle which is similar to the coincidence degree continuation method of Jean Mawhin.
We show that for gradient systems, the homotopy index continuation method may work in cases in which the coincidence degree is zero.
We also give an application of this result to a periodic boundary value problem of second order.

2.1 Sectorial operators generated by partial differential operators

In this section we will state (without proof) a well-known result which shows that many differential operators occurring in elliptic boundary value problems give rise to sectorial operators.
Let $\Omega\subset\mathbb{R}^n$ be a bounded domain of class $C^{2m+\beta}$ where $m\geq 1$ is an integer, and $0<\beta<1$. Consider the following linear differential operators

$$A(x,D) = \sum_{|\alpha|\leq 2m} a_\alpha(x) D^\alpha \tag{1}$$

$$B_j(x,D) = \sum_{|\alpha|\leq m_j} b_\alpha^j(x) D^\alpha \qquad 1\leq j\leq m$$

Here $\alpha=(\alpha_1,\ldots,\alpha_n)$ denotes a multiindex, $|\alpha|=\alpha_1+\ldots+\alpha_n$, $D^\alpha=D_1^{\alpha_1}\ldots D_n^{\alpha_n}$ and $D_j=\partial/\partial x_j$. Moreover, $a_\alpha:\bar{\Omega}\to R$, $|\alpha|\leq 2m$ are Lipschitz continuous on $\bar{\Omega}$.
Furthermore, if $m>1$ then suppose that $\{B_j(x,D)\}_{j=1}^m$ represents Dirichlet boundary conditions i.e.
$B_j=\frac{\partial^{j-1}}{\partial\nu^{j-1}}$ where for $x\in\partial\Omega$, $\nu(x)$ is the outward normal to Ω at x.
If m=1 then assume that

$$B(x,D)u = B_1(x,D)u = \frac{\partial u}{\partial\mu}(x)+h(x)u(x) \tag{2}$$

Here $\mu:\partial\Omega\to R^n$ is a nontangential smoothly varying direction on $\partial\Omega$, $h:\partial\Omega\to R$ is of class $C^1(\partial\Omega)$ and $u:\partial\Omega\to R$ is an arbitrary C^1-function. (2) represents the so-called "mixed-boundary conditions".
Let for $k=1,\ldots,r$, the linear differential operators $(A^k(x,D), B_j^k(x,D))$, $j=1,\ldots,m_k$ be given. Assume that these differential operators satisfy the above properties.
Let $f:\bar{\Omega}\times\mathbb{R}^r\to\mathbb{R}^r$, $(x,u)\to f(x,u)$, $(x,u)\in\bar{\Omega}\times\mathbb{R}^r$ be a continuous mapping. Moreover, assume that f is locally Lipschitzian in $u\in\mathbb{R}^r$, uniformly for $x\in\bar{\Omega}$.

Consider the following system of partial differential equations

$$
\begin{aligned}
&\frac{\partial u_k(t,x)}{\partial t} + A^k(x,D)u_k(t,x) = f_k(x,u(t,x)), \quad x \in \Omega,\ t \in J,\ k=1,\dots,r \\
&B_j^k(x,D)u_k(t,x) = 0 , \qquad x \in \partial\Omega,\ t \in J,\ k=1,\dots,r,\ j=1,\dots,m_k .
\end{aligned} \tag{3}
$$

If J is an interval in $\mathbb{R}$ and $u: J\times\overline{\Omega}\to\mathbb{R}^r$ is a mapping smooth enough so that (3) is satisfied in the classical sense (pointwise), then u is called a <u>classical solution</u> of (3).
If the derivatives in A^k and B_j^k are interpreted in the distributional sense, then we speak of <u>distributional solutions</u>. In most applications, due to regularity theory, distributional solutions are also classical solutions.
Now consider the following system (4):

$$
\begin{aligned}
&A^k(x,D)u_k(x) = f_k(x,u(x)) \quad x \in \Omega,\ k=1,\dots,r \\
&B_j^k(x,D)u_k(x,D) = 0 \qquad x \in \partial\Omega,\ k=1,\dots,r
\end{aligned} \tag{4}
$$

Solutions of (4) are exactly the time independent solutions of (3). We will now concentrate on distributional solutions of (3) and (4) and rewrite these equations in an abstract form. For every $p>1$, let D_p be the closure in $W^{2m,p}(\Omega)$ of the set of all functions $u \in C^{2m}(\overline{\Omega})$ that satisfy the boundary conditions $B_j(x,D)u=0$ on $\partial\Omega (1\leq j\leq m)$.
For $u \in D_p$ define

$$(A_p u)(x) = A(x,D)u(x) \quad . \tag{5}$$

Then $A_p: D_p \to L^p(\Omega)$ is a closed operator.
Now the following result holds:

<u>Theorem 1.1</u> (see e.g. Friedman [1] or Tanabe [1]).

Suppose $A(x,D)$ is strongly elliptic i.e.

$$(-1)^m \sum_{|\alpha|=m} a_\alpha(x)\xi^\alpha > 0 \quad \text{for } x \in \overline{\Omega} \text{ and } \xi \in R^n,\ \xi\neq 0.$$

Then the operator A_p defined by (5) is sectorial on $X_p=L^p(\Omega)$. Moreover, A_p has compact resolvent and the spectrum $\sigma(A_p)$ of A_p is independent of p and consists of a sequence $\{\lambda_k\}$, $k=0,1,2,\dots$ $\mathrm{re}\lambda_0\leq \mathrm{re}\lambda_1\leq \mathrm{re}\lambda_2\leq\dots$, of distinct eigenvalues such that $|\lambda_k|\to\infty$ for

$k\to\infty$. <u>If</u> A_2 <u>is selfadjoint, then these eigenvalues are real</u>.

<u>Remark</u>:

The most common example of a strongly elliptic differential operator is given by $A(x,D)=-\Delta$, where Δ is the Laplace operator. Note that our terminology of "strong ellipticity" follows Friedman [1], while Tanabe [1] calls A "strongly elliptic" if $(-1)^m A$ is strongly elliptic in Friedman's sense. If X_p^α, $\alpha>0$, is the family of fractional power spaces defined by A_p, then by the Sobolev embedding theorems there is a continuous embedding.

$X_p^\alpha \subset C^\nu(\bar{\Omega},\mathbb{R})$ if $0\le\nu<2m\alpha-n/p$.

(see Henry [1], Theorem 1.6.1).

Now given system (3) choose for every k a $p_k>0$. Let $A_{p_k}^k$ be the sectorial operator on X_{p_k} defined by the differential operators $(A^k(x,D), B_j^k(x,0))$ above. Let A be the product of these operators. Then A is sectorial on $X=\prod_{k=1}^{r} X_{p_k}$. (see Henry [1], p. 19).

Moreover the α-th fractional power space of A is the product of the α-th fractional power spaces of the operators $A_{p_k}^k$. Therefore, if we take the numbers $p_k>0$, $k=1,\dots,r$, and $0\le\alpha<1$ in such a way that for every k

$$2m_k\alpha > n/p_k \tag{6}$$

then $X^\alpha \subset C^0(\bar{\Omega},\mathbb{R}^r)$ continuously. In the applications to (3) given in this book, we will always assume that (6) is satisfied.

Now define the operator $\hat{f}:X^\alpha\to X$ by $\hat{f}(u)(x):=f(x,u(x))$, $x\in\bar{\Omega}$.

It is an easy exercise to prove that (by (6)) $\hat{f}$ is a well-defined operator which is Lipschitzian on bounded sets in X^α. $\hat{f}$ is called the <u>Nemitski-operator</u> induced by the mapping f. We will often drop the hat "^" and use the same symbol to denote the mapping f and its Nemitski-operator. With these definitions the system (3) (resp. (4)) can be written abstractly as

$$\frac{du}{dt}+Au = f(u) \qquad \text{for } u\in X^\alpha \tag{7}$$

$$\text{(resp. } Au = f(u) \qquad \text{for } u\in D(A) \tag{8)}$$

(7) is the abstract parabolic equation of the form (S_f) of Section 1.1.

2.2 Center manifolds and their approximation

In this section we recall without proof some basic existence and approximation results for local invariant manifolds near zero with respect to a semilinear parabolic equation. For details the reader is referred to Henry [1], Carr [1], and Chow and Hale [1].

Theorem 2.1 (see Theorem 6.2.1 and Corollary 6.2.2 in Henry [1]).

Suppose A *is sectorial in* X, $0\leq\alpha<1$, U *is a neighborhood of zero in* X^α, $f:U\to X$ *is differentiable,* f' *is Lipschitz continuous and* $f(0)=0$. Let $L=A-f'(0)$ *and assume that* $\sigma(L)\cap\{\lambda|\ \mathrm{re}\ \lambda\leq 0\}$ *is a finite set, isolated in* $\sigma(L)$.
Let $X=X_1\oplus X_2\oplus X_3$ *be the corresponding decomposition of* X *into* L*-invariant subspaces* X_j, $j=1,2,3$ *such that* $\mathrm{re}\ \sigma(L_1)=0$ $\mathrm{re}\ \sigma(L_2)<0$, $\mathrm{re}\ \sigma(L_3)>0$, $(L_j=L_j|X_j,\ j=1,2,3)$, *and let* $\dim(X_1+X_2)<\infty$. *Then there exist open convex neighborhoods* V_j *of zero in* $X_j^\alpha=X^\alpha\cap X_j$, $j=1,2,3$ *such that* $V=V_1\oplus V_2\oplus V_3\subset U$, *and there exist two continuously differentiable mappings* $\Phi:V_1\oplus V_2\to V_3$ *and* $\rho:V_1\to V_2$ *satisfying the following properties:*

$$(1)\quad \Phi'(x_1+x_2)[L_1x_1+L_2x_2-E_1g(x_1+x_2+\Phi(x_1+x_2))-E_2g(x_1+x_2+\Phi(x_1+x_2))]=$$

$$= L_3\Phi(x_1+x_2)-E_3g(x_1+x_2+\Phi(x_1+x_2))$$

$$(2)\quad \rho'(x_1)[L_1x_1-E_1g(x_1+\rho(x_1)+\Phi(x_1+\rho(x_1))]=$$

$$= L_2\rho(x_1)-E_2g(x_1+\rho(x_1)+\Phi(x_1+\rho(x_1)))$$

$$(3)\quad \Phi(x_1+x_2) = 0(\|x_1+x_2\|_\alpha^2) \qquad \text{as } x_1+x_2\to 0 \quad \text{in } X^\alpha$$

$$\rho(x_1) = 0(\|x_1\|_\alpha^2) \qquad \text{as } x_1\to 0 \quad \text{in } X^\alpha$$

Here, $g(u)=f(u)-f'(0)\,u$, *and* E_j *is the projection of* X *onto* X_j, $j=1,2,3$, *induced by the above direct sum decomposition.*

The mapping Φ defines a so-called *local center-unstable manifold*

$$M_{cu}=\{x_1+x_2+\Phi(x_1+x_2)\ |\ x_1\in V_1\ ,\ x_2\in V\ \}.$$

The proof of the existence of Φ follows the same lines as that of Theorem 6.2.1 and Corollary 6.2.2 in Henry [1]. Cf. also Theorem 1 in Chapter 2 of Carr [1] and Theorem 2.11 in Chapter 9 of Chow and Hale [1].

Given the mapping Φ, we consider, on the one hand, the equation

$$\dot{u}+Lu = g(u) \qquad u \in V \tag{1}$$

generating a local semiflow π on V, and, on the other hand, the reduced equation

$$\dot{x}_1+\dot{x}_2+L_1x_1+L_2x_2=E_1g(x_1+x_2+\Phi(x_1+x_2))+E_2g(x_1+x_2+\Phi(x_1+x_2)). \tag{2}$$

Since $X_1 \oplus X_2$ is finite-dimensional, (2) is an ordinary differential equation generating a local semiflow π_Φ on $V_1 \oplus V_2$.

Applying to this reduced equation the center manifold theorem (e.g. Theorem 2.2 in Chapter 9 of Chow and Hale [1]) we obtain the mapping ρ having the desired properties. The set $\{x_1+\rho(x_1) \mid x_1 \in V\}$ is a local center manifold for the reduced equation (2).
We thus obtain the following

Theorem 2.2

If ρ and Φ are as in Theorem 2.1, then define

$$\xi(x_1) = \rho(x_1) \oplus \Phi(x_1+\rho(x_1)) =: x_2 \oplus x_3 .$$

Then $\xi: V_1 \to V_2 \oplus V_3$ is continuously differentiable and

$$\xi'(x_1)[L_1x_1-E_1g(x_1+\xi(x_1))]=L_2x_2+L_3x_3-E_2g(x_1+\xi(x_1))-E_3g(x_1+\xi(x_1)).$$

Proof:

This follows by a simple differentiation using (1) and (2) of Theorem 2.1. ξ defines the set $M_c=\{x_1+\xi(x_1) \mid x_1 \in V_1\}$ called a *local center manifold* for (1).
The following result shows how a local center manifold can be approximated up to any given order of accuracy.

Theorem 2.3

Assume all hypotheses of Theorems 2.1 and 2.2.
Let $\Phi: V_1 \to X_2^\alpha \oplus X_3^\alpha$ be a C^1-mapping with Lipschitzian derivative Φ' and such that $u_1+\Phi(u_1) \in U$ and $\Phi(u_1) \in D(L)$ for $u_1 \in V_1$. Define

$$\Delta(u_1)=\Phi'(u_1)[L_1u_1-E_1g(u_1+\Phi(u_1))]-L\Phi(u_1)+g(u_1+\Phi(u_1))-E_1g(u_1+\Phi(u_1)).$$

Here, E_i is the projection onto X_i in the above direct sum decomposi-

tion. If there exists a $p>1$ *and an* $M>0$ *such that*

$$\|\Delta(u_1)\| \leq M\|u_1\|^p \qquad \text{for all } u_1 \in V_1$$

then, for some $\tilde{M}>0$

$$\|\Phi(u_1)-\xi(u_1)\|_\alpha \leq \tilde{M}\|u_1\|^p \qquad \text{for all } u_1 \in V_1 .$$

Theorem 2.3 is an extension of Theorem 6.2.3 in Henry [1]. The proof is obtained by modifying the arguments from the proof of the latter theorem. Details are omitted.

2.3 The index product formula

Given a local center manifold $M_c=\{x_1+\xi(x_1)\,|\,x_1 \in V_1\}$, where ξ is as in Theorem 2.2 of the last section we can consider the following reduced equation on V_1

$$\dot{x}_1+L_1x_1 = E_1g(x_1+\xi_1(x_1)) \qquad \text{(R)}$$

(R) is the reduction of Eq. (1) of the last section to the center manifold M_c.
From the center manifold theory it is known that all small invariant sets K of Eq. (1) of the last section lie on M_c, i.e. they are governed by the above Eq. (R). Consequently the question arises whether there is a relation between the homotopy index of such an invariant set with respect to the full parabolic equation and the homotopy index of the same set with respect to Eq. (R) on the center manifold. It is the purpose of this section to show that such a relation exists. Since (R) is a finite (usually low) dimensional ordinary differential equation, the index relative to (R) might be easier to compute than the index relative to (1). We will present an example of such a computation in the next section.
We can now state

Theorem 3.1

Let the hypotheses (and, consequently the assertions) of Theorem 2.1 of the last section be satisfied. Let ξ *be as in Theorem 2.2 of the last section and* π_ξ *be the local semiflow on* V_1 *generated by Eq. (R) above.*
Then there exist open and bounded neighborhoods $W_j \subset V_j$ *of zero in*

X_j^α, j=1,2,3 <u>such that the following properties hold</u>:

(1) <u>If</u> $K_\xi \subset W_1$ <u>is isolated</u> (<u>in</u> V_1) <u>and</u> π_ξ-<u>invariant, then</u> $K=\{x_1+\xi(x_1) \mid x_1 \in K_1\}$ <u>is an isolated</u> (<u>in</u> V) <u>and</u> π-<u>invariant set</u>. <u>Moreover, both</u> $h(\pi,K)$ <u>and</u> $h(\pi_\xi,K_\xi)$ <u>are defined and</u>

$$h(\pi,K)=h(\pi_\xi,K_\xi)\wedge\sum^m \tag{1}$$

<u>where</u> m=dim X_2.

(2) If $K \subset W$ <u>is isolated</u> (<u>in</u> V) <u>and</u> π-<u>invariant, then</u> $K_\xi=E_1K\subset W_1$ <u>is isolated</u> (<u>in</u> V_1) <u>and</u> π_ξ-<u>invariant and</u> $K=\{x_1+\xi(x_1) \mid x_1 \in K_\xi\}$. <u>Moreover,</u> $h(\pi,K)$ <u>and</u> $h(\pi_\xi,K_\xi)$ <u>are defined and</u> (1) <u>holds</u>.

Before proceeding to the proof of Theorem 3.1, let us notice the following trivial, but very useful result:

<u>Proposition 3.2</u>

<u>Let</u> X <u>and</u> $\tilde{X}$ <u>be metric spaces and let</u> π (<u>resp</u>. $\tilde{\pi}$) <u>be a local semiflow on</u> X (<u>resp. on</u> $\tilde{X}$).

<u>Suppose that there is a homeomorphism</u> $T:X\to\tilde{X}$ <u>such that for every</u> $x \in X$ <u>and</u> $t \in R^+$, $x\pi t$ <u>is defined if and only if</u> $(T(x))\tilde{\pi}t$ <u>is defined and then</u> $(T(x))\tilde{\pi}t=T(x\pi t)$.

<u>Under this hypothesis, whenever</u> (π,K) <u>is admissible, then</u> $(\tilde{\pi},T(K))$ <u>is admissible and then</u>

$$h(\pi,K)=h(\tilde{\pi},T(K)) \ .$$

The proof of Proposition 3.2 is left to the reader.

A large part of this section will be devoted to the proof of Theorem 3.1.

Let us first notice that by taking appropriate subsets of V_i, if necessary, and substituting them for V_i, we may assume w.l.o.g. that g is bounded on V. Let us further remark that since X_1 and X_2 are finite-dimensional vector spaces it follows that $X_1^\alpha=X_1$ and $X_2^\alpha=X_2$ and the norm on X_1^α, i=1,2 induced by X^α is equivalent to the norm induced by X.

<u>1. Step</u>

We introduce a transformation of variables $T:V\to X_1^\alpha\times X_2^\alpha\times X_3^\alpha$ defined as

$T(u)=T(x_1\oplus x_2\oplus x_3)=(x_1,x_2,x_3-\Phi(x_1+x_2))$.

Let us first prove that T(V) is open in $X_1^\alpha\times X_2^\alpha\times X_3^\alpha$. Let $u^0=x_1^0\oplus x_2^0\oplus x_3^0 \in V$. Then there is an $\varepsilon>0$ such that if $x_i \in X_i^\alpha$ and $\|x_i-x_i^0\|\leq\varepsilon$ for i=1,2,3 then $x_1\oplus x_2\oplus x_3 \in V$. Let $\delta<\varepsilon/2$ be such that if $\|x_i-x_i^0\|_\alpha<\delta$ for i=1,2, then $\|\Phi(x_1+x_2)-\Phi(x_1^0+x_2^0\|_\alpha<\varepsilon/2$. Let $O=\{(y_1,y_2,y_3) \in X_1^\alpha\times X_2^\alpha\times X_3^\alpha \mid \|y_i-y_i^0\|_\alpha<\delta$

for i=1,2,3 }. Here $(y_1^0,y_2^0,y_3^0)=T(u^0)$. Hence $T(u^0)\in O$ and O is open in $X_1^\alpha \times X_2^\alpha \times X_3^\alpha$. We show that $O\subset T(V)$. In fact, let $(y_1,y_2,y_3)\in O$ and define $x_1=y_1$, $x_2=y_2$, $x_3=\Phi(x_1+x_2)+y_3$. Since $x_i^0=y_i^0$ for i=1,2 and $\delta<\varepsilon/2$, we have $x_i\in V_i$, i=1,2. Therefore x_3 is well-defined and

$$\|x_3-x_3^0\|_\alpha \leq \|\Phi(x_1+x_2)-\Phi(x_1^0+x_2^0)\|_\alpha+\|y_3-y_3^0\|_\alpha<\varepsilon/2+\delta<\varepsilon/2+\varepsilon/2 = \varepsilon .$$

Thus $x_3\in V_3$, i.e. $u=x_1+x_2+x_3\in V$. But, obviously, $T(u)=(y_1,y_2,y_3)$ and so $T(V)$ is open.
Let $\tilde{V}=T(V)$. Define $T^{-1}:\tilde{V}\to V$ by $T^{-1}(y_1,y_2,y_3)=(y_1+y_2+(\Phi(y_1+y_2)+y_3))$. It is obvious that T^{-1} is the inverse of T, and so T is a homeomorphism. We denote by (x_1,x_2,y_3) generic points of $\tilde{V}=T(V)$.
Moreover, define the local semiflow $\tilde{\pi}$ on $\tilde{V}$ as the "image" of π under T, i.e. $(Tu)\tilde{\pi}t=T(u\pi t)$. It is easy to see that $\tilde{\pi}$ is the local semiflow on $\tilde{V}$ which is generated by the solutions of system

$$\left.\begin{aligned} \dot{x}_1+L_1x_1 &= N_1(x_1,x_2,y_3)\\ \dot{x}_2+L_2x_2 &= N_2(x_1,x_2,y_3)\\ \dot{y}_3+L_3y_3 &= N_3(x_1,x_2,y_3)\end{aligned}\right\} \tag{2}$$

Here, $N_i(x_1,x_2,y_3)=E_igT^{-1}(x_1,x_2,y_3)$ for i=1,2 , $\tilde{E}_3:=E_1+E_2$ and $N_3(x_1,x_2,y_3)=\Phi'(x_1+x_2)[\tilde{E}_3g(x_1+x_2+\Phi(x_1+x_2))-\tilde{E}_3gT^{-1}(x_1,x_2,y_3)]+$
$+E_3gT^{-1}(x_1,x_2,y_3)-(E_3g(x_1+x_2+\Phi(x_1+x_2))$.

2. Step

Define the transformation

$S:V_1\oplus V_2\to X_1\times X_2$ by $S(x_1\oplus x_2)=(x_1,x_2-\rho(x_1))$.

Proceeding exactly as in the 1st step we prove that $Y=S(V_1\oplus V_2)$ is open in $X_1\times X_2$ and S is a homeomorphism of $V_1\oplus V_2$ onto Y.
We denote by (x_1,y_2) generic points of Y. Let π_Φ be the image of π_Φ under the map S. Then $\tilde{\pi}_\Phi$ is easily seen to be the local semiflow on Y generated by the solutions of the system

$$\left.\begin{aligned} \dot{x}_1+L_1x_1 &= M_1(x_1,y_2)\\ \dot{y}_2+L_2y_2 &= M_2(x_1,y_2)\end{aligned}\right\} \tag{3}$$

where

$$M_1(x_1,y_2) = E_1g(x_1+y_2+\rho(x_1)+\Phi(x_1+y_2+\rho(x_1)))$$

$$\begin{aligned}M_2(x_1,y_2) = {} & \rho'(x_1)[E_1g(x_1+\xi(x_1))- \\ & -E_1g(x_1+y_2+\rho(x_1)+\Phi(x_1+y_2+\rho(x_1)))] + \\ & +E_2g(x_1+y_2+\rho(x_1)+\Phi(x_1+y_2+\rho(x_1))) - \\ & -E_2g(x_1+\xi(x_1)) .\end{aligned}$$

3. Step

Since $\sigma(L)\cap\{\lambda|\mathrm{re}\sigma(\lambda)\leq 0\}$ is a finite set, isolated in $\sigma(L)$, it follows that for some $\gamma>0$, $\mathrm{re}\sigma(L_2)<-\gamma$ and $\mathrm{re}\sigma(L_3)>\gamma$.
Therefore, by Theorem 1.5.3 in Henry [1], there are constants $C,\gamma>0$ such that

$$\|e^{-L_2t}x_2\|\leq Ce^{+\gamma t}\|x_2\| \quad \text{for } x_2\in X_2, \quad t\leq 0 \tag{4}$$

$$\|e^{-L_3t}x_3\|\leq Ce^{-\gamma t}\|x_3\| \quad \text{for } x_3\in X_3, \quad t\geq 0 . \tag{5}$$

Here, $\|\ \|$ is the norm on X.
Define the following homotopies (for $\tau\in[0,1]$):

$$N_j(\tau)(x_1,x_2,y_3) = N_j(x_1,x_2,(1-\tau)y_3),\ j=1,2,3,\ (x_1,x_2,y_3)\in\tilde{V} ,$$

$$M_j(\tau)(x_1,y_2) = M_j(x_1,(1-\tau)y_2) ,\ j=1,2,\ (x_1,y_2)\in Y.$$

By the definition of N_j,M_j and the convexity of V_i, i=1,2,3, it follows that $N_j(\tau)$ (resp. $M_j(\tau)$) is a well-defined locally Lipschitzian mapping from $\tilde{V}$ into X_j, j=1,2,3, (resp. from Y into X_j, j=1,2).

Consider the equations

$$\left.\begin{aligned}\dot{x}_1+L_1x_1 &= N_1(\tau)(x_1,x_2,y_3)\\ \dot{x}_2+L_2x_2 &= N_2(\tau)(x_1,x_2,y_3)\\ \dot{x}_3+L_3y_3 &= N_3(\tau)(x_1,x_2,y_3)\end{aligned}\right\} \tag{6_τ}$$

$$\left.\begin{aligned}\dot{x}_1+L_1x_1 &= M_1(\tau)(x_1,y_2)\\ \dot{y}_2+L_2y_2 &= M_2(\tau)(x_1,y_2)\end{aligned}\right\} \tag{7_τ}$$

We then have the following

Lemma 3.3

There are open balls $\tilde{W}_i$ at zero in X_i^α, $i=1,2,3$ and constants C_1, $\mu>0$ such that $\tilde{W}_i \subset V_i$, $i=1,2,3$, $\tilde{W}=\tilde{W}_1\times\tilde{W}_2\times\tilde{W}_3\subset\tilde{V}$ and such that for all $\tau\in[0,1]$, $t,t_0\in\mathbb{R}$, $t\geq t_0$, the following properties hold:

(1) *If $s\to(x_1(s), x_2(s), y_3(s))$ is a solution of (6_τ) on $[t_0,t]$ lying in $\tilde{W}$ for $s\in[t_0,t]$, then*

$$\|y_3(t)\|_\alpha\leq C_1 e^{-\mu(t-t_0)}\|y_3(t_0)\|_\alpha. \tag{8}$$

(2) *If $s\to(x_1(s), y_2(s))$ is a solution of (7_τ) on $[t_0,t]$ lying in $\tilde{W}_1\times\tilde{W}_2$ for $s\in[t_0,t]$, then*

$$\|y_2(t_0)\|\leq C_1 e^{+\mu(t_0-t)}\|y_2(t)\|. \tag{9}$$

Moreover, if $x_i\in\tilde{W}_i$, $i=1,2$ then $\Phi(x_1+x_2)\in\tilde{W}_3$ and $\rho(x_1)\in\tilde{W}_2$.

Proof of Lemma 3.3:

Since g, $\Phi\in C^1$ and $g(0)=0$, $\Phi(0)=0$, $g'(0)=0$, $\Phi'(0)=0$, it follows that for every $\varepsilon>0$ there is a $\delta(\varepsilon)$, $0<\delta(\varepsilon)<\varepsilon/2$ such that the function $\varepsilon\to\delta(\varepsilon)$ is increasing, $\lim_{\varepsilon\to 0}\delta(\varepsilon)=0$ and such that (writing $B_i(a)=\{x\in X_i^\alpha \mid \|x\|_\alpha<a\}$) :

$$\|\Phi'(x_1+x_2)\|_\alpha\leq\varepsilon \qquad \text{for } x_i\in B_i(\delta(\varepsilon)),\quad i=1,2 \tag{10}$$

$$\|\rho'(x_1)\|\leq\varepsilon \qquad \text{for } x_1\in B_1(\delta(\varepsilon)) \tag{11}$$

$$\|g(x_1+x_2+x_3)-g(\tilde{x}_1+\tilde{x}_2+\tilde{x}_3)\|\leq\varepsilon\sum_{i=1}^{3}\|x_i-\tilde{x}_i\|_\alpha$$
$$\text{for } x_i,\tilde{x}_i\in B_i(\delta(\varepsilon)),\ i=1,2,3. \tag{12}$$

Let $\varepsilon>0$ be fixed and such that $B_i(\varepsilon)\subset V_i$ for $i=1,2,3$. Later on, we will impose additional conditions on ε.
Let $\delta:=\delta(\varepsilon)$. There is a $\delta_1<\delta/2$ such that $\Phi(x_1+x_2)\in B_3(\delta/2)$ and $\rho(x_1)\in B_2(\delta/2)$ for $x_i\in B_i(\delta_1)$, $i=1,2$. It follows that if $x_i\in B_i(\delta_1)$, $i=1,2$, $y_3\in B_i(\delta_1)$, then $x_1+x_2+(1-\tau)y_3\in V$ and from (10) and (12):

$$\begin{aligned}
&\|N_3(\tau)(x_1,x_2,y_3)\|=\|N_3(x_1,x_2,(1-\tau)y_3)\|\leq\\
&\leq\|\Phi'(x_1+x_2)\|_\alpha\cdot\|\tilde{E}_3\|\cdot\|g(x_1+x_2+\Phi(x_1+x_2))-g(x_1+x_2+(1-\tau)y_3+\\
&+\Phi(x_1+x_2))\|+\|E_3\|\cdot\|g(x_1+x_2+(1-\tau)y_3+\Phi(x_1+x_2))-\\
&-g(x_1+x_2+\Phi(x_1+x_2))\|\leq\varepsilon^2\|\tilde{E}_3\|\cdot\|y_3\|_\alpha+\varepsilon\|E_3\|\|y_3\|_\alpha\ .
\end{aligned} \tag{13}$$

Moreover, if $x_1 \in B_1(\delta_1)$ and $y_2 \in B_2(\delta_1)$ then $x_1+(y_2+\rho(x_1)) \in V_1 \oplus V_2$, so $(x_1,y_2) \in Y$, and from (11), (13) we obtain (using the equivalence of the norms $\| \ \|_\alpha$ and $\| \ \|$ on $X_i^\alpha = X_i$, i=1,2 and assuming that $\varepsilon<1$

$$\|M_2(\tau)(x_1,y_2)\| = \|M_2(x_1,(1-\tau)y_2)\| \leq \|\rho'(x_1)\| \|E_1\| \|g(x_1+\rho(x_1)) +$$
$$+\Phi(x_1+\rho(x_1))) - g(x_1+\rho(x_1)+(1-\tau)y_2+\Phi(x_1+(1-\tau)y_2+\rho(x_1)))\| +$$
$$+\|E_2\| \|g(x_1+\rho(x_1)+(1-\tau)y_2+\Phi(x_1+(1-\tau)y_2+\rho(x_1))) - g(x_1+\rho(x_1)+$$
$$+\Phi(x_1+\rho(x_1)))\| \leq 2\varepsilon^2\|E_1\| \|y_2\|_\alpha + 2\varepsilon\|E_2\| \|y_2\|_\alpha \leq (d\varepsilon^2\|E_1\| + d\varepsilon\|E_2\|)\|y_2\| .$$

Here d is such that

$$2\|y\|_\alpha \leq d\|y\| \quad \text{for all } y \in X_1^\alpha + X_2^\alpha .$$

Define $\tilde{W}_i = B_i(\delta_1)$, i=1,2,3.

Now let $s \to (x_1(s), x_2(s), Y_3(s))$, $s \in [t_0,t]$, be a solution of (6_τ) lying in $\tilde{W}$ for all $s \in [t_0,t]$; also let $s \to (x_1(s), y_2(s))$, $s \in [t_0,t]$, be a solution of (7_τ) lying in $\tilde{W}_1 \times \tilde{W}_2$ for $s \in [t_0,t]$.

Then, from the variation-of-constants formula we obtain, using (4), (5) and Theorem 1.4.3 in Henry [1],

$$\|y_3(\tilde{t})\|_\alpha \leq Ce^{-\gamma(\tilde{t}-t_0)}\|y_3(t_0)\|_\alpha +$$
$$+C_\alpha\int_{t_0}^{\tilde{t}} (\tilde{t}-s)^{-\alpha}e^{-\gamma(\tilde{t}-s)}\|N(\tau)(x_1(s), x_2(s), y_3(s))\|ds \leq$$
$$\leq Ce^{-\gamma(\tilde{t}-t_0)}\|y_3(t_0)\|_\alpha + C_\alpha\varepsilon'\int_{t_0}^{\tilde{t}} (\tilde{t}-s)^{-\alpha}e^{-\gamma(\tilde{t}-s)}\|y_3(s)\|_\alpha ds \tag{14}$$

and

$$\|y_2(\tilde{t}_0)\| \leq Ce^{+\gamma(\tilde{t}_0-t)}\|y_2(t)\| + C\int_{\tilde{t}_0}^{t} e^{+\gamma(t_0-s)}\|M_2(\tau)(x_1(s), y_2(s))\|ds \leq$$
$$\leq Ce^{+\gamma(\tilde{t}_0-t)}\|y_2(t)\| + \varepsilon''\int_{\tilde{t}_0}^{t} e^{+\gamma(\tilde{t}_0-s)}\|y_2(s)\|ds \tag{15}$$

Here, $\tilde{t}_0, \tilde{t}$ are arbitrary elements of $[t_0,t]$, $\varepsilon' = \varepsilon^2\|\tilde{E}_3\| + \varepsilon|E_3|$, $\varepsilon'' = d\varepsilon^2\|E_1\| + d\varepsilon\|E_2\|$, and C_α is a constant depending only on L_3 and α.

Now, apply Gronwall's Lemma to (15); moreover, apply Lemma 7.1.1. in Henry [1] to $u(\tilde{t}) = e^{\gamma\tilde{t}}\|y_3(t)\|_\alpha$. Then we obtain constants C_1, $q>0$

such that

$$\|y_3(t)\|_\alpha \leq C_1\|y_3(t_0)\|_\alpha e^{-\gamma(t-t_0)} e^{q(t-t_0)} \tag{16}$$

$$\|y_2(t_0)\| \leq C_1\|y_2(t)\| e^{-\gamma(t-t_0)} e^{\varepsilon''(t-t_0)} \tag{17}$$

Here $C_1=C_1(\alpha,C)$ and $q=(C_\alpha\cdot\varepsilon'\cdot\Gamma(1-\alpha))^{1/1-\alpha}$, where Γ is the gamma function. Choose ε so small that $\gamma-\varepsilon''>\mu>0$ and $\gamma-q>\mu>0$ for some μ.

Then (16) and (17) imply (8) and (9), respectively. The lemma is proved.

Lemma 3.4

Let $\tilde{W}$ be as in Lemma 3.3. Suppose that $\tilde{N}$ (resp. $\tilde{N}_\Phi$) is a closed set in $X_1\times X_2\times X_3^\alpha$ (resp. in $X_1\times X_2$), $\tilde{N}\subset\tilde{W}$ (resp. $\tilde{N}_\Phi\subset\tilde{W}_1\times\tilde{W}_2$), and let $\tilde{\pi}(\tau)$ (resp. $\tilde{\pi}_\Phi(\tau)$) be the local semiflow on $\tilde{V}$ (resp. on $\tilde{V}_1\times\tilde{V}_2$) generated by the solutions of (6_τ) (resp. (7_τ)).
Then $\tilde{N}$ is $\{\tilde{\pi}(\tau_n)\}$-admissible (resp. $\tilde{N}_\Phi$ is $\{\tilde{\pi}_\Phi(\tau_n)\}$-admissible) for every sequence $\{\tau_n\}\subset[0,1]$.
Furthermore, $\tilde{\pi}(\tau)$ does not explode in $\tilde{N}$ (resp. $\tilde{\pi}_\Phi(\tau)$ does not explode in $\tilde{N}_\Phi$) for $\tau\in[0,1]$.
Finally, let $\tilde{K}(\tau)$ (resp. $\tilde{K}_\Phi(\tau)$) be the largest $\tilde{\pi}(\tau)$-invariant set in $\tilde{N}$ (resp. the largest $\tilde{\pi}_\Phi(\tau)$-invariant set in $\tilde{N}_\Phi$). Then $\tilde{K}(\tau)\subset\tilde{W}_1\times\tilde{W}_2\times\{0\}$ (resp. $\tilde{K}_\Phi(\tau)\subset\tilde{W}_1\times\{0\}$).

Proof of Lemma 3.4:

Let $\tilde{N}$ be closed in $X_1\times X_2\times X_3^\alpha$, $\tilde{N}\subset\tilde{W}$, and let $\tilde{N}_\Phi$ be closed in $X_1\times X_2$, $\tilde{N}_\Phi\subset\tilde{W}_1\times\tilde{W}_2$. Take an arbitrary sequence $\{\tau_n\}\subset[0,1]$ and set $\pi_n=\tilde{\pi}(\tau_n)$.

We will first show that $\tilde{N}$ is $\{\pi_n\}$-admissible. In fact, let $\{t_n\}\subset\mathbb{R}^+$ and $\{z_n\}\subset\tilde{N}$ be such that $t_n\to\infty$ and $z_n\pi_n[0,t_n]\subset\tilde{N}$. Writing $z_n\pi_n t=(x_1^n(t),\ x_2^n(t),\ y_3^n(t))$, we obtain from (8):

$$\|y_3^n(t_n)\|_\alpha \leq C_1 e^{-\mu t_n}\|y_3^n(0)\|_\alpha \leq C_1\delta_1 e^{-\mu t_n} .$$

It follows that $\|y_3^n(t_n)\|_\alpha\to 0$. Moreover, since $X_1\times X_2$ is finite-dimensional and both $\{x_1^n(t_n)\}$ and $\{x_2^n(t_n)\}$ are bounded, it follows that there is a subsequence of $\{(x_1^n(t_n),\ x_2^n(t_n))\}$, denoted by the same symbol $\{(x_1^n(t_n),\ x_2^n(t_n))\}$, which converges to some $(x_1,x_2)\in X_1\times X_2$. Consequently, $z_n\tilde{\pi}t_n\to(x_1,x_2,0)$ and as $\tilde{N}$ is closed in $X_1\times X_2\times X_3^\alpha$, we have $(x_1,x_2,0)\in\tilde{N}$. This proves that $\tilde{N}$ is $\{\pi(\tau_n)\}$-admissible. That $\tilde{N}_\Phi$

is $\{\pi_\Phi(\tau_n)\}$-admissible follows trivially from the fact that $X_1 \times X_2$ is finite-dimensional and $\tilde{N}_\Phi$ is bounded.

The second assertion of the lemma follows immediately from Theorem I.2.4.

Now let $\tilde{K}(\tau)$ be the largest $\tilde{\pi}(\tau)$-invariant set in $\tilde{N}$, and let $\tilde{K}_\Phi(\tau)$ be the largest $\tilde{\pi}_\Phi(\tau)$-invariant set in $\tilde{N}_\Phi$. We will first prove that $\tilde{K}(\tau) \subset \tilde{W}_1 \times \tilde{W}_2 \times \{0\}$. In fact, if $z \in \tilde{K}(\tau)$ then there is a full solution $t \to \sigma(t)$ of (6_τ) such that $\sigma(0)=z$ and $\sigma(t) \subset \tilde{K}(\tau) \subset \tilde{N}$ for all $t \in \mathbb{R}$. Writing $\sigma(t)=(x_1(t), x_2(t), y_3(t))$ we obtain from Lemma 3.3 for every $t_0, t \in \mathbb{R}$, $t_0<t$

$$\|y_3(t)\|_\alpha \le C_1 e^{-\mu(t-t_0)} \|y_3(t_0)\|_\alpha \le C_1 \delta_1 e^{-\mu(t-t_0)} .$$

Letting $t_0 \to -\infty$ we see that $\|y_3(t)\|_\alpha=0$ for all $t \in \mathbb{R}$, thus proving $\tilde{K}(\tau) \subset \tilde{W}_1 \times \tilde{W}_2 \times \{0\}$.

Similarly, let $t \to \hat{\sigma}(t)$ be a full solution of (7_τ) such that $\hat{\sigma}(t) = (x_1(t), y_2(t)) \in \tilde{K}_\Phi(\tau)$ for $t \in \mathbb{R}$. Taking t_0, t arbitrary, $t_0<t$, we obtain from (9)

$$\|y_2(t_0)\| \le C_1 e^{+\mu(t_0-t)} \|y_2(t)\| \le C_1 \delta_1 e^{+\mu(t_0-t)} .$$

Taking $t \to +\infty$ we obtain $\|y_2(t_0)\|=0$. Hence $\tilde{K}_\Phi(\tau) \subset \tilde{W}_1 \times \{0\}$, as claimed. The lemma is proved.

4. Step

We can now complete the proof of Theorem 3.1:

Choose W_i to be small open balls at zero in X_i^α for which $\mathrm{Cl}W_i \subset V_i$, $i=1,2,3$, such that $T(\mathrm{Cl}W_1 \oplus \mathrm{Cl}W_2 \oplus \mathrm{Cl}W_3) \subset \tilde{W}$ and $S(\mathrm{Cl}W_1 \oplus \mathrm{Cl}W_2) \subset \tilde{W}_1 \times \tilde{W}_2$, and such that $\Phi(W_1 \oplus W_2) \subset W_3$ and $\rho(W_1) \subset W_2$.

Let $W = W_1 \oplus W_2 \oplus W_3$.

We first prove (1) of Theorem 3.1:

Let $K_\xi \subset W_1$ be an isolated (in V_1), π_ξ-invariant set. Then, by Theorem 2.2. it is easily seen that

$$K = \{x_1 + \rho(x_1) + \Phi(x_1 + \rho(x_1)) \mid x_1 \in K_\xi\}$$

is π-invariant.

Being isolated in V_1, K_ξ is, by definition, closed in V_1. But since the closure $\mathrm{Cl}W_1$ of W_1 in X_1 is in V_1, it follows that K_ξ is closed in X_1. Hence there exists a set B, closed in X_1, $B \subset W_1$, such that B is an isolating (in V_1) neighborhood of K_ξ.

Let $N=B\oplus ClW_2\oplus ClW_3$. Then N is closed in X^α, and, by our assumptions, K is in the interior of N relative to the topology of X^α. Hence K is in the interior of N relative to V. To prove that N is an isolating neighborhood of K, it therefore suffices to show that K=K', where K' is the largest π-invariant set in N. Obviously $K\subset K'$. By the definition of T and $\tilde{\pi}$ it follows that $\tilde{N}=T(N)$ is closed in $X_1\times X_2\times X_3^\alpha$, $\tilde{K}'=T(K')$ is a $\tilde{\pi}$-invariant set and $\tilde{N}\subset\tilde{W}$. Now use Lemma 3.4: Since $\tilde{\pi}(0)=\tilde{\pi}$, it follows that $\tilde{K}'=\tilde{K}'(0)\subset\tilde{W}_1\times\tilde{W}_2\times\{0\}$. Thus whenever $x_1+x_2+x_3\in K'$, then $x_3=\Phi(x_1+x_2)$. Therefore $K_\Phi'=\{x_1+x_2 \mid x_1+x_2+\Phi(x_1+x_2)\in K'\}$ is π_Φ-invariant and $K_\Phi'\subset B\oplus ClW_2$. Obviously $S(B\oplus ClW_2)=\tilde{N}_\Phi$ is closed in $X_1\times X_2$ and $\tilde{N}_\Phi\subset\tilde{W}_1\times\tilde{W}_2$. Again from Lemma 3.4 we conclude that $\tilde{K}_\Phi'=S(K_\Phi')\subset W_1\times\{0\}$. For every $x_1+x_2\in K_\Phi'$ it follows that $x_2=\rho(x_1)$ and consequently $K_\xi'=\{x_1 \mid x_1+\rho(x_1)+\Phi(x_1+\rho(x_1))\in K'\}$ is π_ξ-invariant. But since $K_\xi'\subset B$, it follows that $K_\xi'=K_\xi$. It follows that K'=K and this proves that N is an isolating neighborhood of K.

Let $\tilde{K}(\tau)$ be the largest $\tilde{\pi}(\tau)$-invariant set in $\tilde{N}=T(N)$. By Proposition 3.2 $\tilde{K}(0)=\tilde{K}=T(K)$. By Lemma 3.4 $\tilde{K}(\tau)\subset\tilde{W}_1\times\tilde{W}_2\times\{0\}$ and so $\tilde{K}(\tau)$ is easily seen to be independent of τ, i.e. $\tilde{K}(\tau)=T(K)=\tilde{K}$ for every $\tau\in[0,1]$. Therefore $\tilde{N}$ is an isolating neighborhood of $\tilde{K}(\tau)$, for every $\tau\in[0,1]$. Using this fact, Lemma 3.4 and Theorem I.2.4 it is trivially seen that (1) and (2) of Definition I.12.1 are satisfied for the map $\psi:\tau\to(\tilde{\pi}(\tau),\ \tilde{K}(\tau))$. Hence ψ is S-continuous. Theorem I.12.2 implies that

$$h(\tilde{\pi}(0),\ \tilde{K}(0)) = h(\tilde{\pi}(1),\ \tilde{K}(1)). \tag{18}$$

Let $K_\Phi=\{x_1+\rho(x_1) \mid x_1\in K_\xi\}$ and $\hat{K}_1=\{(x_1,\rho(x_1)) \mid x_1\in K_\xi\}$. Then, obviously, $\tilde{K}=\hat{K}_1\times\{0\}$. Moreover, it is immediate that $\tilde{\pi}(1)=\hat{\pi}_\Phi\times\pi_3$, where $\hat{\pi}_\Phi$ is the image of π_Φ under the homeomorphism $x_1\oplus x_2\to(x_1,x_2)$, and π_3 is the local semiflow on V_3 generated by the solutions of

$$\dot{y}_3+L_3y_3 = 0 .$$

Using (5) we conclude from Theorem I.11.1 that $h(\pi_3,\{0\})=\Sigma^0$. Hence by Theorem I.10.5

$$h(\tilde{\pi}(1),\ \hat{K}_1\times\{0\}) = h(\hat{\pi}_\Phi,\hat{K}_1)\wedge\Sigma^0 = h(\hat{\pi}_\Phi,\hat{K}_1) \tag{19}$$

Here we used the obvious fact that $(Y,y_0)\wedge(S^0,s_0)$ is homeomorphic to (Y,y_0), for every pointed space (Y,y_0).

Using Proposition 3.2 we obtain from (18) and (19)

$$h(\pi,K)=h(\tilde{\pi},\tilde{K})=h(\tilde{\pi}(0),\tilde{K}(0))=h(\tilde{\pi}(1),\tilde{K}(0))=h(\hat{\pi}_\Phi,\hat{K}_1)=h(\pi_\Phi,K_\Phi). \quad (20)$$

Consider now the map $\tilde{\psi}:\tau\to(\tilde{\pi}_\Phi(\tau),\tilde{K}_\Phi(\tau))$ where $\tilde{K}_\Phi(\tau)$ is the largest $\tilde{\pi}_\Phi(\tau)$-invariant set in $\tilde{N}_\Phi=S(B\oplus\mathrm{Cl}W_2)$. As before, it follows that $\tilde{K}_\Phi(\tau)\subset\tilde{W}_1\times\{0\}$ and therefore $\tilde{K}_\Phi(\tau)$ is independent of τ, i.e. $\tilde{K}_\Phi(\tau)\equiv\tilde{K}_\Phi(0)=\tilde{K}_\Phi=S(K_\Phi)=K_\xi\times\{0\}$. It follows again that $\tilde{\psi}$ is an S-continuous mapping. Theorem I.12.2 implies

$$h(\tilde{\pi}_\Phi(0),\ \tilde{K}_\Phi) = h(\tilde{\pi}_\Phi(1),\ \tilde{K}_\Phi) \ . \quad (21)$$

But $\tilde{\pi}_\Phi(1)=\pi_\xi\times\pi_2$, where π_2 is the local semiflow on V_2 generated by the solutions of

$$\dot{y}_2+L_2y_2 = 0 \ .$$

Using (4) we conclude from Theorem I.11.1 that $h(\pi_2,\{0\})=\Sigma^m$ where $m=\dim X_2$.
Now we obtain from (20), (21), Proposition 3.2, and Theorem I.10.5

$$h(\pi_\Phi,K_\Phi)=h(\tilde{\pi}_\Phi,\tilde{K}_\Phi)=h(\tilde{\pi}_\Phi(0),\tilde{K}_\Phi)=h(\tilde{\pi}_\Phi(1),\tilde{K}_\Phi)=h(\pi_\xi,K_\xi)\wedge\Sigma^m. \quad (22)$$

(20) and (22) imply

$$h(\pi,K) = h(\pi_\xi,K_\xi)\wedge\Sigma^m \ ,$$

i.e. formula (1).
Part (1) of Theorem 3.1 is proved.
To prove part (2), let $K\subset W$ be an isolated (in V) and π-invariant set. Thus, by definition, K is closed in V. But since $\mathrm{Cl}W\subset V$, it follows that K is closed in X^α. Hence there is a set N, closed in X^α, $N\subset\mathrm{Cl}W$, such that N is an isolating neighborhood of K. Let $\tilde{N}=T(N)$ and $\tilde{K}=T(K)$. Using the same arguments as those in the proof of part (1), we prove that for every $x_1+x_2+x_3\in K$, it follows that $x_2=\rho(x_1)$ and $x_3=\Phi(x_1+\rho(x_1))$, i.e. $x_1+x_2+x_3=x_1+\xi(x_1)$. It follows that $K_\xi=E_1K$ is a π_ξ-invariant set, isolated in V_1, and $K=\{x_1+\xi(x_1)\mid x_1\in K_\xi\}$. Therefore the rest of part (2) follows from part (1) and the theorem is proved.
If $X_1=\{0\}$ in Theorem 3.1, then 0 is a <u>hyperbolic</u> equilibrium, i.e. a so-called <u>saddle-point property</u> is satisfied (see e.g. Henry [1]). Then $h(\pi_\xi,\{0\})$ is obviously equal Σ^0. Using the formula

$\sum^m \wedge \sum^n = \sum^{m+n}$, $m,n \geq 0$ we obtain that

$$h(\pi,\{0\}) = \sum^m .$$

If A has compact resolvent, then this latter result is valid under more general assumptions on f:

Theorem 3.5

Suppose that A is sectorial in X and has compact resolvent. Let $0 \leq \alpha < 1$ and assume that U is a neighborhood of zero in X^α, $f:U \to X$, is locally Lipschitzian, $f(0)=0$ and $f'(0)$ exists. Let $L=A-f'(0)$ and assume that there is a decomposition $X=X_2 \oplus X_3$ into L-invariant subspaces X_i, $i=2,3$ with $\operatorname{re}\sigma(L_2)<-\delta<0$ and $\operatorname{re}\sigma(L_3)>\delta>0$ for some $\delta>0$ and $L_i:=L|X_i$, $i=2,3$.
Then $\dim X_2=:m<\infty$. Moreover, $K=\{0\}$ is an isolated invariant set, $h(\pi,\{0\})$ is defined and $h(\pi,\{0\}=\sum^m$.

Proof:

We may assume that U is a ball at zero of radius $\rho>0$, and that f is Lipschitzian and (hence) bounded on U. Define $g_\tau:U \to X$ to be

$$g_\tau(u) = (1-\tau)(f(u)-f'(0)\,u), \qquad \tau \in [0,1],$$

and π_τ to be the local semiflow on U generated by

$$\dot{u}+Lu = g_\tau(u) \qquad (S_\tau) .$$

We will prove that the map $\Phi(\tau)=(\pi_\tau,\{0\})$ is a well-defined S-continuous map from $[0,1]$ into $S=S(U)$. Assuming this for the moment, we infer from Theorem I.12.2 that $h(\pi_0,\{0\})=h(\pi_1,\{0\})$. However, (S_0) is our original equation, so $\pi_0=\pi$. Furthermore, (S_1) is a linear equation to which Theorem I.11.1 applies. By that theorem $h(\pi_1\{0\})=\sum^m$. This proves Theorem 3.5, except for our claim that Φ is S-continuous. To prove our claim, it is only necessary to show that there is a closed set $N \subset U$, such that for every $\tau \in [0,1]$, N is an isolating neighborhood of $K=\{0\}$, relative to π_τ. Then Theorems I.2.4 and I.4.4 immediately imply all hypotheses of Definition I.12.1, i.e. Φ is S-continuous. If such a set N does not exist then there is a sequence $\tau_n \in [0,1]$ and for every n, a full solution $u_n:\mathbb{R} \to U$ of $\pi_n:=\pi_{\tau_n}$ with

$$c_n = \sup_{t \in \mathbb{R}} \|u_n(t)\|_\alpha \neq 0 , \qquad c_n < \|u_n(0)\|_\alpha + 1$$

and $c_n \to 0$ as $n \to \infty$. Let $g_n = g_{\tau_n}$.
Then for all n large enough, the maps $\tilde{g}_n : V \to X$, $\tilde{g}_n(v) = c_n^{-1} g_n(c_n v)$ are well-defined and Lipschitzian, hence, bounded on V, where $V = \{u \in X^\alpha \mid \|u\|_\alpha < 2\}$. Let $v_n(t) = c_n^{-1} u_n(t)$. Then v_n is a full solution of $\tilde{\pi}_n$ where $\tilde{\pi}_n$ is the local semiflow generated on V by solutions of

$$\dot{v} + Lv = \tilde{g}_n(v) \qquad (\tilde{S}_n) .$$

Let $N = \{v \in X^\alpha \mid \|v\|_\alpha \leq 1\}$. Then Theorem I.4.4 implies that N is strongly $\{\tilde{\pi}_n\}$-admissible. In particular, $\{v_n(0)\}_{n \geq 1}$ contains a convergent subsequence. We may assume that $v_n(0) \to v_0$ as $n \to \infty$. Now $v_n(0) \in A^-_{\tilde{\pi}_n}(N)$ for all n. Since $\tilde{g}_n \to 0$ as $n \to \infty$ uniformly on V, $\tilde{\pi}_n \to \tilde{\pi}_0$ where $\tilde{\pi}_0$ is the restriction to V of the linear semiflow π_1 above.
Now Theorems I.2.4 and I.4.5 imply that $v_0 \in A_{\pi_1}(N) \cap \partial N$. However, by Theorem I.11.1 $A_{\pi_1}(N) = \{0\}$, a contradiction.
This completes the proof of Theorem 3.5.

Remark:

The decomposition $X = X_2 \oplus X_3$ with the properties listed in Theorem 3.5 exists if and only if $\sigma(L) = \sigma_2 \cup \sigma_3$ where re $\sigma_2 < -\delta < 0$ and re $\sigma_3 > \delta > 0$. Moreover, m is necessarily the total algebraic multiplicity of eigenvalues λ of L with re $\lambda < 0$. (cf. Corollary I.11.2 and Henry [1]).

2.4 A one-dimensional example

As an application of our result in the last section, we will now compute the homotopy index of $K = \{0\}$ for a one-dimensional Dirichlet boundary value problem.
More precisely, we consider the following equation

$$\left.\begin{aligned} \frac{\partial u}{\partial t} &= \frac{\partial^2 u}{\partial x^2} + f(u(t,x)) , \qquad x \in (0,\pi) \\ u(t,0) &= u(t,\pi) = 0 . \end{aligned}\right\} \tag{1}$$

Here $f : \mathbb{R} \to \mathbb{R}$ is a locally Lipschitzian function.
Let us write equation (1) in the form (S_f). To this end, let $X = L^2(0,\pi)$, $D(A) = H^2(0,\pi) \cap H_0^1(0,\pi)$,

$$A : D(A) \to X , \qquad Au = -\frac{\partial^2 u}{\partial x^2}$$

where $\frac{\partial}{\partial x}$ is the derivative in the distributional sense. A is a sectorial operator with compact resolvent and $\sigma(A)=\{(r+1)^2 \mid r=0,1,2,...\}$ with corresponding (normalized) eigenfunctions $e_r(x)=\sqrt{2/\pi}\sin(r+1)x$, $x \in (0,\pi)$. Write $\lambda_r=(r+1)^2$, $r\geq 0$.
As in section 2.1, it follows that f defines a corresponding locally Lipschitzian Nemitski operator $\hat{f}:X^{1/2}\to X$. Also, it is well-known that $X^{1/2}=H_0^1(0,\pi)$.
It follows that we can apply our theory to equation (1). First assume the nonresonance case, i.e. suppose that $\mu=f'(0)$ exists and $\mu \notin \sigma(A)$, say $\lambda_r<\mu<\lambda_{r+1}$, where $r\geq -1$ and $\lambda_{-1}:=-\infty$. Then $\sigma(A-f'(0))=\sigma(A)-\mu$. Therefore there are exactly r+1 negative eigenvalues of $\sigma(A-f'(0))$ and all the other eigenvalues are positive. All eigenvalues being simple, Theorem 3.5 implies that $h(\pi,\{0\})$ is defined and

$$h(\pi,\{0\}) = \Sigma^{r+1} .$$

We now consider the <u>resonance</u> case. Assume f to have a Lipschitzian derivative.near zero and let $\mu=f'(0)=\lambda_r$ for some $r\geq 0$. We shall compute the index of $\{0\}$ under some (generic) hypotheses on f. More precisely, let

$$f(s) = \lambda_r s+as^K+\beta(s) \tag{2}$$

where $a\neq 0$, $K\geq 2$ is an integer, and $\beta(s)=0(s^{K+1})$ for $s\to 0$.
Let $g(s):=f(s)-\lambda_r s$ and $L=A-\lambda_r$. Let $X_1=\text{span}\{e_r\}$, $X_2=\text{span}\{e_0,...,e_{r-1}\}$, $X_3=\overline{\text{span}\{e_i \mid i\geq r+1\}}$. Then $X=X_1\oplus X_2\oplus X_3$, the spaces X_i are mutually orthogonal and L-invariant, and writing $L_i=L|X_i$, we see that $L_1=0$, i.e.

$$\sigma(L_1) = 0 \quad , \quad \sigma(L_2) < 0 \ , \quad \text{and } \sigma(L_3) > 0 .$$

Hence there is a one-dimensional local center manifold close to zero which can be described by a mapping $\xi:V_1\to V_2\oplus V_3$ where V_i is a neighborhood of zero in $X_i^{1/2}$, for i=1,2,3.
Set $\Phi(u_1)=\langle u_1,e_r\rangle^K\cdot v$, $u_1 \in X_1$ with $v\perp e_r$, $v \in D(A)$ to be determined later.
Then for $x \in (0,\pi)$

$$\begin{aligned} g(u_1+\Phi(u_1))(x) &= a[\langle u_1,e_r\rangle e_r(x)+\langle u_1,e_r\rangle^K v(x)]^K + \\ &+\beta(\langle u_1,e_r\rangle e_r(x)+\langle u_1,e_r\rangle^K v(x)) \end{aligned} \tag{3}$$

Since $|\langle u_1,e_r\rangle|$ is an equivalent norm on X_1, it follows that

$$g(u_1+\Phi(u_1)) = a<u_1,e_r>^K e_r^K+0(\|u_1\|^{K+1}) . \tag{4}$$

Computing $\Delta(u_1)$ by means of the formula in Theorem 2.3 we obtain

$$\begin{aligned}\Delta(u_1) &= K<u_1,e_r>^{K-1}[-a<u_1,e_r>^K<e_r^K,e_r>v] - \\ &\quad -<u_1,e_r>^K Lv+a<u_1,e_r>^K e_r^K - \\ &\quad -a<u_1,e_r>^K<e_r^K,e_r>e_r+0(\|u_1\|^{K+1}) = \\ &= -<u_1,e_r>^K\{Lv-ae_r^K+a<e_r^K,e_r>e_r\}+0(\|u_1\|^{K+1})\end{aligned} \tag{5}$$

Let us now choose $v\perp e_r$ in such a way that the expression in braces is zero. Since $L=A-\lambda_r$ and $<ae_r^K-a<e_r^K,e_r>e_r,e_r>=0$, such a v can be determined in a unique way.

Using (5) and Theorem 2.3 we therefore conclude that $\|\xi(u_1)-\Phi(u_1)\|_{X^{1/2}}\leq\tilde{M}\|u_1\|^{K+1}$ for all $u_1\in V_1$. Thus the reduced equation on the center manifold reads

$$\dot{u}_1 = E_1 g(u_1+\xi(u_1)) = a\cdot\delta<u_1,e_r>^K\cdot e_r+0(\|u_1\|^{K+1}) \tag{6}$$

Here $\delta=<e_r^K,e_r>=(\sqrt{2/\pi})^{K+1}\int_0^\pi \sin^{K+1}(r+1)x\,dx$.

Writing $y=<u_1,e_r>$, we obtain from (6) an equivalent scalar equation

$$y = a\cdot\delta y^K+0(y^{K+1}) =: h(y). \tag{7}$$

If K is odd or if K is even and r is even, then $\delta>0$. Hence is this case,

$$\operatorname{sign} h(y) = \operatorname{sign} ay^K$$

for small $y\neq0$.

This implies that the index $h_\xi(\pi,\{0\})$ is given by the index of $\{0\}$ with respect to the scalar equation

$$y = ay^K . \tag{8}$$

However, this latter index is trivial to compute and thus we obtain the following

Proposition 4.1

Let $f(s)=\lambda_r s+as^K+0(s^{K+1})$ as $s\to 0$, with $a\neq 0$ and $K\geq 2$. Then if K is odd or if K is even and r is even, then $\{0\}$ is an isolated invariant set of π_f.
Furthermore, the index $h(\pi_\xi,\{0\})$ on the center manifold is given by

$$h(\pi_\xi,\{0\}) = \begin{cases} \overline{0} & \text{if K is even and r is even,} \\ \Sigma^1 & \text{if K is odd and } a>0 , \\ \Sigma^0 & \text{if K is odd and } a<0 . \end{cases}$$

The index of $\{0\}$ with respect to the full semiflow π_f is given by

$$h(\pi_f,\{0\}) = \begin{cases} \overline{0} & \text{if K is even and r is even,} \\ \Sigma^{r+1} & \text{if K is odd and } a>0 , \\ \Sigma^r & \text{if K is odd and } a<0 . \end{cases}$$

Proof:

The formula for $h(\pi_\xi,\{0\})$ is the result of the preceding remarks; the formula for $h(\pi_f,\{0\})$ follows from the index product formula (Theorem 3.1)

$$h(\pi_f,\{0\}) = h(\pi_\xi,\{0\}) \wedge \Sigma^m$$

by noticing that $m=r$.
We now consider the critical case K even, r odd. Here it seems possible that the index depends on the higher order terms in the expansion of f. To show this, we will restrict ourselves to a special case

$$f(s) = \lambda_r s+as^K+bs^{2K-1}+0(s^{2K}) \text{ as } s\to 0 . \tag{9}$$

Here K is even, r is odd, $a\neq 0$ and b is arbitrary. Define $g(s)=f(s)-\lambda_r s$.
Let $\Phi(u_1):=\langle u_1,e_r\rangle^K v+\langle u_1,e_r\rangle^{2K-1}w$ where v is as before and $w\perp e_r$ is to be determined later.
It follows for $x\in(0,\pi)$,

$$g(u_1+\Phi(u_1))(x)=a(u_1(x)+\langle u_1,e_r\rangle^K v(x)+\langle u_1,e_r\rangle^{2K-1}w(x))^K+b(u_1(x)+\langle u_1,e_r\rangle^K v(x)+\langle u_1,e_r\rangle^{2K-1}w(x))^{2K-1}+0(\|u_1\|^{2K})=a(\langle u_1,e_r\rangle^K e_r^K(x)+K\langle u_1,e_r\rangle^{2K-1}e_r^{K-1}(x)v(x))+b\langle u_1,e_r\rangle^{2K-1}e_r^{2K-1}(x)+0(\|u_1\|^{2K}) .$$

Computing $\Delta(u_1)$ we obtain therefore

$$\Delta(u_1) = -\langle u_1,e_r\rangle^{2K-1}\{Lw-aKe_r^{K-1}v+aK\langle e_r^K,v\rangle e_r-be_r^{2K-1} + b\langle e_r^{2K-1},e_r\rangle e_r\}+0(\|u_1\|^{2K}).$$

As before, $w\perp e_r$ can be found such that the expression in braces is zero.

Theorem 2.3 implies that

$$\|\Phi(u_1)-\xi(u_1)\|_{X^{1/2}} = 0(\|u_1\|^{2K}) \quad \text{as } u_1\to 0 .$$

Consequently, the reduced equation on the center manifold reads

$$\dot{u}_1 = \langle u_1,e_r\rangle^{2K-1}(aK\langle e_r^K,v\rangle+b\langle e_r^{2K-1},e_r\rangle)e_r+0(\|u_1\|^{2K}) . \tag{10}$$

Writing $y=\langle u_1,e_r\rangle$, we obtain an equivalent scalar equation

$$\dot{y} = \{aK\langle e_r^K,v\rangle+b\langle e_r^{2K-1},e_r\rangle\}y^{2K-1}+0(y^{2K}) . \tag{11}$$

If the expression in braces is $\neq 0$, then the index of $\{0\}$ with respect to (11) exists and is equal to the index of $\{0\}$ with respect to

$$\dot{y} = \eta y^{2K-1} \tag{12}$$

where $\eta=aK\langle e_r^K,v\rangle+b\langle e_r^{2K-1},e_r\rangle$.

Since v is independent of b and $\langle e_r^{2K-1},e_r\rangle>0$, this index does, in fact, depend on b.

Now, it is easily seen that (with $\mu:=r+1$)

$$v(x)=C_0\frac{1}{\mu}\cos\mu x\int_0^x \sin^{K+1}\mu s\,ds-C_0\frac{1}{\mu^2(K+1)}\cdot\sin^{K+2}\mu x,$$

$$C_0 := a(\sqrt{2/\pi})^K .$$

Now, a simple integration yields

$$\langle v,e_r^K\rangle = -aC_1$$

where $C_1>0$ is a constant.

This shows that the first term in η is negative, and we obtain the following

Proposition 4.2

Let K be even and r be odd. Moreover, let f be as in (9) above. Then the index of $\{0\}$ with respect to π_f exists if either $b\leq 0$ or else if b is positive and large enough.
Furthermore,

$$h(\pi_f,\{0\}) = \begin{cases} \Sigma^r & \text{if } b\leq 0 \text{ in } (9)\ , \\ \Sigma^{r+1} & \text{if } b \text{ is positive and large enough}\ . \end{cases}$$

2.5 Asymptotically linear systems

If X is a normed space and π is a local semiflow on X, then by K_∞ we will denote the union of all full bounded orbits of π. In other words, a point $x\in X$ is in K_∞ if and only if there is a full solution $\sigma:\mathbb{R}\to X$ of π, $\sigma(0)=x$, and $\sigma[\mathbb{R}]$ is bounded. K_∞ is obviously an invariant set, and, in general, K_∞ is unbounded. There are situations, however, where K_∞ is bounded. This is so, for example, if π is dissipative in some sense (a concept which will be defined later on). In such a case, K_∞ is compact and connected and attracts all solutions, i.e. every solution tends to K_∞ as $t\to\infty$.
In this section we will encounter another class of semiflows for which K_∞ is bounded (and in fact compact) but K_∞ is not an attractor, in general. This is so for asymptotically linear systems with non-resonance at infinity.
More precisely, we have the following

Theorem 5.1

Suppose that A is sectorial in X and has compact resolvent. Let $0\leq\alpha<1$ and assume that $f:X^\alpha\to X$ is locally Lipschitzian, mapping bounded sets in X^α into bounded sets in X and asymptotically linear, i.e. such that there exists a bounded linear map $B:X^\alpha\to X$ such that $(f(u)-Bu)/\|u\|_\alpha\to 0$ as $\|u\|_\alpha\to\infty$.
Let $L=A-B$ and assume that there exists a decomposition $X=X_2\oplus X_3$ into L-invariant subspaces X_i, $i=2,3$, with $\mathrm{re}\,\sigma(L_2)<-\delta<0$ and $\mathrm{re}\,\sigma(L_3)>\delta>0$ for some $\delta>0$ and $L_i:=L|X_i$, $i=2,3$. Then $\dim X_2=:m<\infty$. Moreover, if π is the local semiflow on X^α generated by the solutions of

$$\dot{u}+Au = f(u) \tag{1}$$

and K_∞ is the union of all full bounded orbits of π, then $(\pi,K_\infty)\in\mathcal{S}(X^\alpha)$

and $h(\pi,K_\infty)=\Sigma^m$.

Proof:

This theorem is "dual" to Theorem 3.5 and has a "dual" proof. We therefore only sketch the details.
Define $g_\tau:X^\alpha\to X$, $\tau\in[0,1]$ to be the map

$$g_\tau(u) = (1-\tau)(f(u)-Bu)$$

and let π_τ be the local semiflow on X^α generated by the solutions of

$$\dot{u}+Lu = g_\tau(u) \qquad (S_\tau)$$

Let K_τ be the union of all full bounded orbits of π_τ. We will show that the map $\Phi:[0,1]\to S$, $\Phi(\tau)=(\pi_\tau,K_\tau)$ is well-defined and S-continuous. Assuming this and using the fact that $K_1=\{0\}$ by Theorem I.11.1 we immediately get the assertion of the theorem.
To prove that Φ is well-defined and S-continuous, it is only necessary to show that there is a bounded set $N\subset X^\alpha$ such that $K_\tau\subset N$ for all $\tau\in[0,1]$. Suppose this is not true.
Then there is a sequence $\tau_n\in[0,1]$ and a sequence of full bounded solutions $t\to u_n(t)$ of $\pi_n:=\pi_{\tau_n}$ with

$$c_n := \sup\|u_n(t)\|_\alpha\to\infty \quad \text{as } n\to\infty$$

and $\|u_n(0)\|>c_n-1$.
Let $v_n(t)=c_n^{-1}u_n(t)$ and $g_n:X^\alpha\to X$ be defined as $\tilde{g}_n(v)=c_n^{-1}g_{\tau_n}(c_nv)$, $n\in\mathbb{N}$,
Then $\tilde{g}_n$ is locally Lipschitzian for $n\in\mathbb{N}$. We will show that for every $\rho\geq 0$,

$$\sup_{\|v\|_\alpha\leq\rho}\|\tilde{g}_n(v)\|\to 0 \quad \text{as } n\to\infty .$$

Let $\varepsilon>0$. Then there is an r such that if $\|u\|_\alpha>r$ then $\|f(u)-Bu\|\leq\varepsilon\|u\|_\alpha$. Moreover, there is an $M=M(r)\geq\|B\|\cdot r$ such that if $\|u\|_\alpha\leq r$ then $\|f(u)\|\leq M$. It follows that

$$\|\tilde{g}_n(v)\| = (1-\tau_n)c_n^{-1}\|f(c_nv)-B(c_nv)\|$$

and so

$$\|\tilde{g}_n(v)\| \leq 2Mc_n^{-1} \quad \text{if } c_n\cdot\|v\|_\alpha\leq r$$

and

$$\|\tilde{g}_n(v)\| \leq \varepsilon\|v\|_\alpha \leq \varepsilon\rho \qquad \text{if } c_n\cdot\|v\|_\alpha > r .$$

This implies the desired claim.
Now proceeding exactly as in the proof of Theorem 3.5 we get that a subsequence of v_n converges to $v\equiv 0$. However, this contradicts our assumption that

$$\|v_n(0)\| > 1-c_n^{-1}\to 1 \qquad \text{as } n\to\infty.$$

The theorem is proved.
The assumptions of Theorem 3.5 are in particular satisfied for systems of partial differential equations discussed in Section 2.1, if we assume that the function $f:\overline{\Omega}\times\mathbb{R}^r\to\mathbb{R}^r$ has the property that

$$(f(x,s)-\lambda^* s)/\|s\| \to 0 \tag{2}$$

as $\|s\|\to\infty$, $s\in\mathbb{R}^r$ uniformly for $x\in\overline{\Omega}$ and $\lambda^*\notin\sigma(A)$. In other words, the nonlinearity is asymptotically linear with nonresonant slope.
If the operator A is self-adjoint (as it is often the case in the applications) and if $f(x,s)$ is the gradient of a function $F(x,s)$, i.e. $f(x,s)=\text{grad}_s F(x,s)$, then Theorem 3.5 may be improved. Before discussing such an improvement we need the following important concepts.

Definition 5.2

Let X be a metric space and π be a local semiflow on X. A point $x_0\in X$ is called an equilibrium of π, if the constant function $\sigma(t)\equiv x_0$, $t\in\mathbb{R}$, is a solution of π.
A continuous function $V:X\to\mathbb{R}$ is called a Liapunov-function for π, if for every $x\in X$, the function $t\to V(x\pi t)$ is nonincreasing for $t\in[0,\omega_x)$. π is called gradient-like with respect to V, if V is a Liapunov-function for π and whenever σ is a nonconstant full solution of π, then $t\to V(\sigma(t))$ is not a constant function.
The following proposition holds:

Proposition 5.3

Let $V:X\to\mathbb{R}$ be a Liapunov function for π. If $J=\mathbb{R}^+$ (resp. $J=\mathbb{R}^-$) and $\sigma:J\to X$ is a solution of π with $\sigma[J]$ relatively compact, then V is constant on $\omega(\sigma)$ (resp. on $\omega^*(\sigma)$). If in addition, π is gradient-like with respect to V, then $\omega(\sigma)$ (resp. $\omega^*(\sigma)$) contains only equilibria of π.

The proof is a trivial exercise using the fact that $\omega(\sigma)$ (resp. $\omega^*(\sigma)$) are invariant sets.
As an obvious corollary we obtain

Proposition 5.4

If π is gradient-like with respect to V and $\sigma:\mathbb{R}\to X$ is a nonconstant full solution of π with $\sigma[\mathbb{R}]$ relatively compact, then $\omega(\sigma)$ and $\omega^*(\sigma)$ are nonempty disjoint sets containing only equilibria of π.
Consequently, a nonconstant full solution σ of a gradient-like local semiflow with $\sigma[\mathbb{R}]$ compact joins two disjoint sets of equilibria. Such a solution σ is also known as a *heteroclinic orbit*.
We now have the following

Theorem 5.5

Assume the following hypotheses:

(1) $(H,\langle\ ,\ \rangle)$ is a Hilbert space;
$(X,\|\ \|)$ is a real Banach space;
X is an $\mathbb{R}$-subspace of H, and the inclusion $X\subset H$ is continuous.

(2) $\tilde{D}\subset\tilde{H}$; $\tilde{A}:\tilde{D}\to H$ is a self-adjoint linear operator on H bounded below, $D\subset\tilde{D}\cap X$, $\tilde{A}[D]\subset X$, and $A:=\tilde{A}|D$, $A:D\to X$, is a sectorial operator in X, defining the fractional power spaces $(X^\alpha,\|\ \|_\alpha)$. $0\le\alpha\le 1$, $X^0=X$, $X^1=D$. Moreover, A has compact resolvent.

(3) $1/2\le\alpha<1$; $f_m:X^\alpha\to X$, $m\in\mathbb{N}$, is a sequence of locally Lipschitzian mappings, and $G_m:X^\alpha\to\mathbb{R}$, $m\in\mathbb{N}$, is a sequence of Fréchet-differentiable mappings, such that:

(3.1) for every $u,h\in X^\alpha$, $m\in\mathbb{N}$

$$DG_m(u)(h) = \langle f_m(u), h\rangle$$

(3.2) for some $M>0$ and all $m\in\mathbb{N}$, $u\in X^\alpha$

$$\|f_m(u)\| \le M(\|u\|_\alpha+1)$$

(3.3) for some $\nu,\delta>0$, and all $m\in\mathbb{N}$, $u\in D$

$$\|-Au+f_m(u)\|_H \ge \nu\|u\|_H-\delta$$

where $\|\cdot\|_H$ is the norm of H.
Under all these hypotheses, there exists an $L>0$ such that

<u>for every</u> $m \in \mathbb{N}$, <u>and every full bounded solution</u> $t \to u(t)$ <u>of</u>

$$\dot{u} + Au = f_m(u) \qquad (4_m)$$

$$\sup_{t \in \mathbb{R}} \|u(t)\|_\alpha \leq L .$$

<u>Proof</u>:

Let π_m be the local semiflow on X^α generated by (4_m).

<u>1. Step</u>:

Since $\tilde{A}$ is self-adjoint and bounded below, it follows that for some $k>0$, $\tilde{A}+kI \geq 0$ and re $\sigma(A+kI) > \delta_0 > 0$. Write $\tilde{A}_1 = \tilde{A}+kI$, $A_1 = A+kI$. It follows that $\tilde{A}_1^{1/2}$ is well-defined, $\tilde{A}_1^{1/2}u = A_1^{1/2}u$ for $u \in X^{1/2}$ and since $\alpha \geq 1/2$, i.e. $X^\alpha \subset X^{1/2}$ it follows that the map $X^\alpha \ni u \to \tilde{A}_1^{1/2} u \in H$ is a bounded linear map.

Define $V_m : X^\alpha \to \mathbb{R}$ as

$$V_m(u) = \frac{1}{2}(\langle \tilde{A}_1^{1/2}u, \tilde{A}_1^{1/2}u \rangle - k\langle u,u \rangle) - G_m(u) .$$

If $t \to u(t) \in X^\alpha$ is differentiable for $t \in (t_1, t_2)$, then

$$\frac{dV_m(u(t))}{dt} = \langle \tilde{A}_1^{1/2}u(t), \tilde{A}_1^{1/2}\dot{u}(t) \rangle - k\langle u(t), \dot{u}(t) \rangle - \\ - \langle f_m(u(t)), \dot{u}(t) \rangle. \qquad (5)$$

Hence, if $t \to u(t)$ is a solution of π_m, $u(t) \in D$, for $t \in (t_1, t_2)$, then (5) implies

$$\frac{dV_m(u(t))}{dt} = \langle (A+kI)u, -Au+f_m(u) \rangle - \langle ku, -Au+f_m(u) \rangle - \langle f_m(u), -Au+f_m(u) \rangle \\ = -\langle Au+f_m(u), -Au+f_m(u) \rangle = -\|-Au+f_m(u)\|_H^2 \qquad (6)$$

It follows that π_m is gradient-like with respect to V_m.

<u>2. Step</u>

We claim that there is an $L_0 > 0$ such that if $m \in \mathbb{N}$ and u is an equilibrium of π_m, then $\|u\|_\alpha < L_0$.
In fact, if this is not true, we can assume w.l.o.g. that there is a sequence $\{u_n\}$ of equilibria of π_{m_n} such that $c_n = \|u_n\|_\alpha \to \infty$.

Let $v_n = c_n^{-1} u_n$.

Let $\tilde{f}_n : X^\alpha \to X$ be defined as $\tilde{f}_n(v) = c_n^{-1} f_{m_n}(c_n v)$. Let $\tilde{\pi}_n$ be the local semiflow generated by $\dot{v} + Av = \tilde{f}_n(v)$.

By hypothesis (3.2), $\|\tilde{f}_n(v)\| \le M(\|v\|_\alpha + c_n^{-1})$.

Since $Au_n = f_{m_n}(u_n)$, i.e. $(A+kI)u_n = f_{m_n}(u_n) + ku_n$ it follows that $(A+kI)v_n = \tilde{f}_n(v_n) + kv_n$ i.e. $v_n = (A+kI)^{-1}(\tilde{f}_n(v_n) + kv_n)$. Since A has compact resolvent, we conclude that, w.l.o.g., v_n converges in X to some $v \in X^\alpha$. Hence $\|v_n - v\|_H \to 0$ as $n \to \infty$. By hypothesis (3.3), $0 \ge \nu \|v_n\|_H - \delta \cdot c_n^{-1}$, so $0 \ge \nu \|v\|_H$, i.e. $v = 0$, a contradiction since $\|v_n\|_\alpha = 1$. This proves our claim.

3. Step

Suppose now that the theorem is not true.

Then there is, w.l.o.g., a sequence $t \to u_n(t)$, of full bounded solutions of (π_{m_n}) such that

$$c_n = \sup_{t \in \mathbb{R}} \|u_n(t)\|_\alpha \to \infty \qquad \text{as } n \to \infty$$

and

$$\|u_n(0)\|_\alpha \ge c_n - 1 > 0 .$$

Let $v_n(t) = c_n^{-1} u_n(t)$. Let $\tilde{f}_n : X^\alpha \to X$ be defined as above, i.e.

$\tilde{f}_n(v) = c_n^{-1} f_n(c_n v)$.

Notice that in hypothesis (3) we can assume w.l.o.g. $G_m(0) \equiv 0$, since, otherwise, we can replace G_m by $G_m - G_m(0)$.

Define $\tilde{G}_n : X^\alpha \to \mathbb{R}$, $\tilde{G}_n(v) = c_n^{-2} \cdot G_{m_n}(c_n v)$.

It is easily seen that $D\tilde{G}_n(v)h = \langle \tilde{f}_n(v), h \rangle$ for $v, h \in X^\alpha$. Therefore, arguing as in Step 1, we see that $\tilde{\pi}_n$ is gradient-like with respect to $\tilde{V}_n : X^\alpha \to \mathbb{R}$,

$$\tilde{V}_n(v) = \frac{1}{2}(\langle A_1^{1/2} v, A_1^{1/2} v \rangle - k \langle v, v \rangle) - \tilde{G}_n(v) .$$

Moreover, since $G_m(v) = \int_0^1 DG_m(t \cdot v)(v)\,dt = \int_0^1 \langle f_m(t \cdot v), v \rangle dt$, we obtain, with some constant $C > 0$

$$|G_m(v)| \le \int_0^1 \|f_m(tv)\|_H \|v\|_H dt \le C \int_0^1 \|f_m(tv)\| \|v\|_\alpha dt \le CM\|v\|_\alpha(\|v\|_\alpha + 1) .$$

Now let $\tilde{u}_n$ be an equilibrium of π_{m_n}. Then, by Step 2 $\|\tilde{u}_n\|_\alpha \le L_0$ and, for $\tilde{v}_n := c_n^{-1}\tilde{u}_n$, we have

$$|\tilde{V}_n(\tilde{v}_n)| \le \frac{1}{2}c_n^{-2}(\|A_1^{1/2}\tilde{u}_n\|_H^2 + k\|\tilde{u}_n\|_H^2) + c_n^{-2}|G_{m_n}(\tilde{u}_n)| \le c_n^{-1}\tilde{C} \tag{7}$$

with some constant $\tilde{C}>0$, independent of n and of the equilibrium $\tilde{u}_n$. Since $t \to v_n(t)$ is a full bounded solution of $\tilde{\pi}_n$, its α- and ω-limit sets contain only equilibria of $\tilde{\pi}_n$, i.e. elements $\tilde{v}_n = c_n^{-1} \cdot \tilde{u}_n$, where $\tilde{u}_n$ is an equilibrium of π_{m_n}.

Therefore, for every $t_1 < t_2$ we obtain from (7)

$$0 \le \tilde{V}_n(v_n(t_1)) - \tilde{V}_n(v_n(t_2)) \le 2c_n^{-1}\tilde{C} . \tag{8}$$

However, by hypothesis (3.3),

$$\begin{aligned} \tilde{V}_n(v_n(t_1)) - \tilde{V}_n(v_n(t_2)) &= \int_{t_2}^{t_1} \frac{d}{dt}(\tilde{V}_n(v_n(t)))dt = \\ &= \int_{t_1}^{t_2} \|-Av_n(t) + \tilde{f}_n(v_n(t))\|_H^2 dt \ge \\ &\ge \int_{t_1}^{t_2} \|\dot{v}_n(t)\|_H (\nu\|v_n(t)\|_H - \delta \cdot c_n^{-1})dt \ge \\ &\ge \beta_n \cdot \|v_n(t_2) - v_n(t_1)\|_H , \end{aligned} \tag{9}$$

where $\beta_n := \inf_{t \in [t_1, t_2]} (\nu\|v_n(t)\|_H - \delta \cdot c_n^{-1})$.

Hence by (8) and (9)

$$\|v_n(t_2) - v_n(t_1)\|_H \le 2c_n^{-1}\beta_n^{-1}\tilde{C} , \qquad \text{if } \beta_n > 0 .$$

Fix $\varepsilon > 0$.

If $\|v_n(t)\|_H \ge \nu^{-1}(c_n^{-1}\delta + \varepsilon)$ for $t \in [t_1, t_2]$ then $\beta_n \le \varepsilon$, therefore

$$\|v_n(t_2) - v_n(t_1)\|_H \le 2\varepsilon\tilde{C}c_n^{-1} . \tag{10}$$

Since the α-limit set of v_n consists of equilibria $\tilde{v}$ of $\tilde{\pi}_n$, satisfying, by (3.3), the property

$$0 \ge \nu\|\tilde{v}\|_H - \delta c_n^{-1} , \quad \text{i.e. } \|\tilde{v}\|_H \le \nu^{-1}c_n^{-1}\delta ,$$

It follows that whenever $\|v(t_2)\|_H > \nu^{-1}(c_n^{-1}\delta+\varepsilon)$ for some $t_2 \in \mathbb{R}$, there is a $t_1 < t_2$ such that $\|v(t_1)\|_H = \nu^{-1}(c_n^{-1}\delta+\varepsilon)$ and $\|v(t)\|_H \geq \nu^{-1}(c_n^{-1}\delta+\varepsilon)$ for $t \in [t_1,t_2]$.
Hence, by (10)

$$\|v_n(t_2)\|_H \leq \|v_n(t_2)-v_n(t_1)\|_H + \|v_n(t_1)\|_H \leq 2\varepsilon\tilde{C}c_n^{-1} + \nu^{-1}(c_n^{-1}\delta+\varepsilon) .$$

It follows that

$$\|v_n(t)\|_H \leq 2\varepsilon\tilde{C}c_n^{-1} + \nu^{-1}(c_n^{-1}\delta+\varepsilon) \quad \text{for } t \in \mathbb{R},\ n \in \mathbb{N} . \tag{11}$$

Now let $N=\{v \in X^\alpha \mid \|v\|_\alpha \leq 1\}$.
By hypothesis (3.2) there is a $B>0$ such that $\|\tilde{f}_n(v)\| \leq B$ for all $n \in \mathbb{N}$ and all $v \in N$. Therefore Theorem I.4.4 implies that N is strongly $\{\tilde{\pi}_n\}$-admissible.
It follows that a subsequence of $\{v_n(0)\}$, again denoted by $\{v_n(0)\}$ converges in X^α to some point $w_0 \in X^\alpha$ with $\|w_0\|_\alpha = 1$. However, (11) implies that $\|w_0\|_H = 0$, i.e. $w_0 = 0$, a contradiction which proves the theorem.
As a corollary to the proofs of Theorems 3.5 and 5.5 we obtain the following

Theorem 5.6

*Assume hypotheses (1) and (2) of Theorem 5.5. Moreover, let $1/2 \leq \alpha < 1$ and U be a neighborhood of 0 in X^α. Let $f: U \to X$ be locally Lipschitzian and assume that $f(0)=0$ and that $f'(0)$ exists and $f'(0)=0$.
Suppose there is a Fréchet-differentiable map $G: U \to \mathbb{R}$ such that for every $u \in U$ and $h \in X^\alpha$,*

$$DG(u)(h) = \langle f(u), h \rangle .$$

Consider the equation

$$Au = \lambda u + f(u) \qquad (T_\lambda)$$

*for an arbitrary $\lambda \in \mathbb{R}$.
Then every point $(\lambda^*,0) \in \mathbb{R} \times U$ with $\lambda^* \in \sigma(A)$ is a bifurcation point of (T_λ), i.e. there exists a sequence $(\lambda_n,u_n) \in \mathbb{R} \times U$ such that $(\lambda_n,u_n) \to (\lambda^*,0)$ as $n \to \infty$, $u_n \neq 0$ and u_n is a solution of (T_{λ_n}) for $n \in \mathbb{N}$
Moreover, every point $(\lambda^*,0)$ with $\lambda^* \notin \sigma(A)$ is not a bifurcation point.*

Proof:

Let $\lambda \in \mathbb{R}$ be arbitrary. Then $u_0 \in U$ is a solution of (T_λ) if and only if u_0 is an equilibrium of the local semiflow π_λ on U generated by equation

$$\dot{u}(t)+Au = \lambda u+f(u) \qquad (P_\lambda) .$$

Let now $\lambda^* \notin \sigma(A)$. Then, proceeding as in the proof of Theorem 3.5 one can easily show that there is an $\varepsilon>0$ and the neighborhood N of 0 in X^α, $N \subset U$, such that $K_\lambda=\{0\}$ is the largest π_λ-invariant set in N for all $\lambda \in (\lambda^*-\varepsilon,\lambda^*+\varepsilon)$. This proves that $(\lambda^*,0)$ is not a bifurcation point of (T_λ), establishing the second part of our assertion.
Now let $\lambda^* \in \sigma(A)$. Suppose $(\lambda^*,0)$ is not a bifurcation point. Then, there exist an $\varepsilon>0$ and a closed bounded neighborhood N of 0 in X^α, $\dot{N} \subset U$, such that whenever $u_0 \in N$ is a solution of (T_λ), $\lambda \in (\lambda^*-\varepsilon,\lambda^*+\varepsilon)$, then $u_0=0$.
Now define $V_\lambda: U \to \mathbb{R}$, $\lambda \in \mathbb{R}$ as

$$V_\lambda(u) = \frac{1}{2}(<A_1^{1/2}u,\ A_1^{1/2}u>-k<u,u>)-\lambda<u,u>-G(u) .$$

As in the proof of Theorem 5.5 (see equations (5), (6) in that proof) it follows that π_λ is gradient-like with respect to V_λ. Now let $\lambda \in (\lambda^*-\varepsilon,\lambda^*+\varepsilon)$ and $t \to u(t)$ be a full solution of π_λ contained in N. Then $u[\mathbb{R}]$ is relatively compact. If $u(t) \equiv u_0$, i.e. if u_0 is an equilibrium of π_λ, then by our assumption, $u(t) \equiv u_0=0$. If $t \to u(t)$ is not a constant solution, then, by Proposition 5.4 $\omega^*(u)$ and $\omega(u)$ are disjoint, nonempty and contain only equilibria of π_λ. Since $\omega^*(u) \cup \omega(u) \subset N$, it follows that $\omega^*(u)=\omega(u)=\{0\}$, a contradiction.
This argument proves that N is an isolating neighborhood of $K_\lambda=\{0\}$ relative to π_λ, for $\lambda \in (\lambda^*-\varepsilon,\lambda^*+\varepsilon)$. It is therefore clear that the map $\Phi: \lambda \to (\pi_\lambda,\{0\})$, $\lambda \in (\lambda^*-\varepsilon,\lambda^*+\varepsilon)$ is well-defined and $\mathcal{S}$-continuous. Therefore Theorem I.12.1 implies that $h(\pi_\lambda,\{0\})=$const. for $\lambda \in (\lambda^*-\varepsilon,\lambda^*+\varepsilon)$. We may assume that $(\lambda^*-\varepsilon,\lambda^*+\varepsilon) \cap \sigma(A)=\{\lambda^*\}$. Let $\lambda_1=\lambda^*-\varepsilon/2$, $\lambda_2=\lambda^*+\varepsilon/2$.
Then Theorem 3.5 and the remark following the proof of that theorem imply that

$$h(\pi_{\lambda_i},\{0\}) = \Sigma^{m_i} \quad , \ i=1,2$$

where m_i is the total algebraic multiplicity of all eigenvalues $\lambda \in \sigma(A)$ with $\lambda<\lambda_i$, $i=1,2$.
But m_2 is m_1 plus the algebraic multiplicity of λ^*. Hence $m_1<m_2$ and so $\sum^{m_1} \neq \sum^{m_2}$, a contradiction.
The theorem is proved.
Let us now recall that a global semiflow π on a normed space X is called _point-dissipative_, if there exists a bounded set $B \subset X$ with the property that for every $x \in X$ there is a $t_0=t_0(x)$ such that $x\pi t \in B$ for $t \geq t_0$. π is called _conditionally completely continuous_ for $t \geq t_0$, if for any bounded set $B \subset X$ there is a compact set $B^* \subset X$ such that for any $t \geq t_0$ and any $x \in X$ for which $x\pi[0,t] \subset B$, it follows that $x\pi[t_0,t] \subset B^*$. Note that if π is conditionally completely continuous for $t \geq t_0$ with some $t_0>0$, then every bounded set $N \subset X$ is strongly π-admissible. By well-known results on dissipative systems, if π is point-dissipative and conditionally completely continuous for $t \geq t_0$ with some $t_0>0$, then the union J of all full bounded orbits of π is a compact and connected global attractor (see e.g. the proofs of Theorem 3.1 and Lemma 3.3 in Chapter 4 of Hale [1]).
Here, by a _global attractor_ we mean that for every $x \in X$, $\mathrm{dist}(x\pi t,J) \to 0$ as $t \to \infty$.
Let A be sectorial on X with compact support and $f:X^\alpha \to X$, $0 \leq \alpha<1$, be locally Lipschitzian and map bounded sets in X^α into bounded sets in X. Then, by Theorem I.4.3, the local semiflow π generated by $\dot{u}+Au=f(u)$ is conditionally completely continuous for $t \geq t_0$ with $t_0>0$ arbitrary.
We now have the following

Theorem 5.7

Assume hypotheses (1) and (2) of Theorem 5.5. Furthermore, suppose that $\tilde{A}$ has compact resolvent. Let λ_i, $i \geq 0$ be the eigenvalues of A and λ_{-1} be an arbitrary (finite) number $<\lambda_0$. Suppose $1/2 \leq \alpha<1$ and that $f:X^\alpha \to X$ is a locally Lipschitzian mapping satisfying the following condition:

(E) _there are $\varepsilon,\delta>0$ and $k \geq -1$ such that for all $u \in X^\alpha$_

$$\|f(u)-\lambda^* u\|_H \leq \omega \|u\|_H+\delta$$

where $\lambda^*=\frac{1}{2}(\lambda_k+\lambda_{k+1})$ and $\omega=\frac{1}{2}(\lambda_{k+1}-\lambda_k)-\varepsilon$.

Let m^* be the total algebraic multiplicity of eigenvalues λ of A with $\lambda \leq \lambda_k$.

Suppose also that there is a Fréchet-differentiable mapping $G: X^{\alpha} \to \mathbb{R}$ *such that for all* $u, h \in X^{\alpha}$

$$DG(u)\cdot(h) = \langle f(u), h\rangle .$$

Finally, let π *be the local semiflow generated by solution of*

$$\dot{u} + Au = f(u)$$

and K_{∞} *be the union of all full bounded orbits of* π. *Then* $(\pi, K_{\infty}) \in S$ *and*

$$h(\pi, K_{\infty}) = \Sigma^{m^*} .$$

Furthermore π *is a global semiflow, and if* $k>1$ *(i.e. if* $m \neq 0$*) then is not point dissipative and* K_{∞} *has a nonempty unstable manifold.*

Proof:

It is an immediate consequence of condition (E) and Corollary 3.3.5 in Henry [1] that π is a global semiflow.
Define the following homotopy

$$g_{\tau}(u) = (1-\tau)f(u) + \tau\lambda^* u \quad \text{for } \tau \in [0,1].$$

Let π_{τ} be the semiflow generated by

$$\dot{u} + Au = g_{\tau}(u)$$

and K_{τ} be the union of all full bounded orbits of π.
We will prove that the map $\Phi(\tau) = (\pi_{\tau}, K_{\tau})$ is well-defined and S-continuous. In fact, for this to be true, it is only necessary that there exist a common bound (in X^{α}) for all sets $K_{\tau}, \tau \in [0,1]$. Suppose that such a common bound does not exist. Then there is a sequence $\tau_m \in [0,1]$ and a sequence u_m of full bounded solutions of π_{τ_m} with $c_m = \sup_{t \in \mathbb{R}} \| u_m(t) \|_{\alpha} \to \infty$ as $m \to \infty$.
Let $f_m := g_{\tau_m}$ and $G_m: X^{\alpha} \to \mathbb{R}$ be defined as $G_m(u) = (1-\tau_m)G(u) + (1/2) . \tau_m \lambda^* \langle u,u\rangle$.
Then hypotheses (1), (2)aand (3.1), (3.2) of Theorem 5.5 are clearly satisfied.
By our assumption, $\lambda^* \in \rho(A)$ and hence $(\tilde{A}-\lambda^*)^{-1}: H \to H$ is a Hermitian operator. Hence $\| (A-\lambda^*)^{-1} \| \leq \max\{ |\mu| \; | \mu \in \sigma((\tilde{A}-\lambda^*)^{-1})\}$. Since $(\tilde{A}-\lambda^*)^{-1}$

is compact, by hypothesis (1), it follows that whenever $\mu\in\sigma(\tilde{A}-\lambda^*)^{-1}$, $\mu\neq 0$, then μ is an eigenvalue of $(\tilde{A}-\lambda^*)^{-1}$, hence $1/\mu$ is an eigenvalue of $A-\lambda^*$.
This implies for every $u\in D$

$$\|u\|_H = \|(A-\lambda^*)^{-1}(A-\lambda^*)u\|_H \leq \max\{|\mu| \; |\mu\in\sigma((\tilde{A}-\lambda^*)^{-1})\}\|(\tilde{A}-\lambda^*)u\|_H \leq$$

$$\leq \max\{|\lambda|^{-1}|\lambda\in\sigma(\tilde{A}-\lambda^*)\}\;\|(\tilde{A}-\lambda^*)u\|_H .$$

Hence

$$\|(A-\lambda^*)u\|_H \geq C_0\|u\|_H$$

where $C_0=\min\{|\lambda| \; |\lambda\in\sigma(\tilde{A}-\lambda^*)\}$.
Now $\mu\in\sigma(\tilde{A}-\lambda^*)$ if and only if $\lambda^*+\mu\in\sigma(\tilde{A})$ if and only if $\mu=\lambda_j-\frac{1}{2}(\lambda_k+\lambda_{k+1})$ for some $j\geq 0$.
Therefore $|\mu|\geq\frac{1}{2}(\lambda_{k+1}-\lambda_k)$ and so $C_0>\omega$.
Now we obtain for all $u\in D$ and $m\in\mathbb{N}$

$$\|-Au+f_m(u)\|_H=\|-Au+\lambda^*u-\lambda^*u+f_m(u)\|_H\geq\|-Au+\lambda^*u\|_H-\|f_m(u)-\lambda^*u\|_H =$$

$$=\|-Au+\lambda^*u\|_H-(1-\tau_m)\|f(u)-\lambda^*u\| \geq$$

$$\geq (C_0-\omega)\|u\|_H-\delta .$$

This proves hypothesis (3.3) and so Theorem 5.5 immediately leads to a contradiction. Hence, indeed, the map $\tau\to(\pi_\tau,K_\tau)$ is well-defined and S-continuous. Since $(\pi_0,K_0)=(\pi,K_\infty)$, it follows that

$$h(\pi,K_\infty) = h(\pi_1,K_1) = h(\pi_1,\{0\}) = \Sigma^{m^*}$$

by Corollary I.11.2.
We shall now prove the last assertion of the theorem. Let $m>0$. Then Σ^m is the homotopy type of a connected set. Therefore an application of Corollary I.11.9 yields the existence of a full solution $t\to u^*(t)$ with $\sup_{t\leq 0}\|u^*(t)\|_\alpha<\infty$ and $\sup_{t\geq 0}\|u^*(t)\|=\infty$.
This shows on the one hand that K_∞ is not an attractor for the solution u^* as $t\to\infty$. In particular, π cannot be point-dissipative. On the other hand, $\omega^*(u^*)$ is bounded and invariant, so $\omega^*(u^*)\subset K_\infty$. Therefore

the solution u^* lies on the unstable manifold of K_∞.
The proof is complete.
As a simple consequence we obtain

Corollary 5.8

Under the assumptions of Theorem 5.7, there exists a solution $u_0 \in D(A)$ *of*

$$Au = f(u) . \tag{12}$$

Proof:

In fact, $h(\pi,K_\infty)=\Sigma^{m^*}\neq\bar{0}$. It follows that $K_\infty\neq\emptyset$. Let $v_0 \in K_\infty$. Then either v_0 is an equilibrium of π, i.e. a solution of (12), or else there exists a nonconstant full bounded solution $t\to u(t)$ with $u(0)=v_0$. By Proposition 5.4 $\omega(u)$ and $\omega^*(u)$ are nonempty and contain only equilibria, i.e. solutions of (12).
The proof is complete.

We shall now apply Theorem 5.7 and Corollary 5.8 to parabolic equations considered in section 1 of this chapter. We will only treat the case $r=1$, i.e. a single equation, leaving it to the reader to formulate and prove the corresponding results for systems $(r>1)$.

Theorem 5.9

Consider the equations (1.3) *and* (1.4) *with* $r=1$. *Assume that the linear differential operators in* (1.3) *and* (1.4) *satisfy the conditions listed in section* 2.1.
Let A_p *be as in* (1.5) *and assume that* A_2 *is self-adjoint.*
Let $\lambda_0<\lambda_1<\ldots$ *be the (common) eigenvalues of the operators* A_p *and* λ_{-1} *be an arbitrary (finite) number* $<\lambda_0$.
Let the nonlinearity $f^1=f:\bar{\Omega}\times\mathbb{R}\to\mathbb{R}$ *in* (1.3) *and* (1.4) *be continuous and locally Lipschitzian in* $s\in\mathbb{R}$, *uniformly for* $x\in\bar{\Omega}$.
Suppose that f *satisfies the following condition* (E*):

(E*) *There are* $\rho>0$, $\varepsilon>0$ *and* $k\geq-1$ *such that whenever* $x\in\bar{\Omega}$ *and* $s\in\mathbb{R}$ *with* $|s|\geq\rho$, *then*

$$\lambda_k+\varepsilon \leq f(x,s)/s \leq \lambda_{k+1}-\varepsilon .$$

Consider the abstract versions (1.7) *(resp.* 1.8) *of* (1.3) *(resp.* (1.4)) *where* $p>n$, $1/2\leq\alpha<1$, $X=L^p(\Omega)$, $H=L^2(\Omega)$, $A=A_p$, $\tilde{A}=A_2$ *and* $f:X^\alpha\to X$ *is the Nemitski operator defined by the function* f.

Let π be the local semiflow on X^α generated by the solution of (1.7). Then the following properties hold:

(1) _π is a global semiflow._

(2) _if K_∞ is the union of all full bounded orbits of π, then K_∞ is compact, $h(\pi,K_\infty)$ is defined and_

$$h(\pi,K_\infty) = \Sigma^{m^*}$$

where m^* is the total algebraic multiplicity of all eigenvalues λ of A with $\lambda \leq \lambda_k$.

Moreover:

(2.1) _if $k=-1$ (i.e. if $m^*=0$), then π is point-dissipative._

(2.2) _if $k>-1$ (i.e. if $m^*>0$), then π is not point-dissipative and K_∞ has a nonempty unstable manifold._

(3) _There exists at least one (classical) solution of equation (1.8)._

Proof:

Let $G:\bar{\Omega}\times\mathbb{R}\to\mathbb{R}$ be defined as

$$G(x,s) = \int_0^s f(x,t)\,dt \tag{13}$$

and $\hat{G}:X^\alpha\to\mathbb{R}$ be the corresponding Nemitski operator, i.e.

$$\hat{G}(u)(x) = G(x,u(x)) \ .$$

Using the imbedding $X^\alpha \subset C^0(\bar{\Omega})$ it is an easy exercise to show that all hypotheses of Theorem 5.7 are satisfied with G, and f replaced by $\hat{G}$ resp. $\hat{f}$ in that theorem.

Therefore Theorem 5.7 implies properties (1), (2) and (2.2). Moreover, Corollary 5.8 and regularity theory implies property (3). It therefore remains to prove property (2.1).

Let $k=-1$.

We use arguments from the proof of Theorem 5.7. Since $\tilde{A}$ is self-adjoint and $\lambda^*<\lambda_0$, it follows that $\sigma(A-\lambda^*)\geq C_0>0$ where C_0
$C_0=\min\{|\lambda| \ | \lambda\in\sigma(\tilde{A}-\lambda^*)\}\geq\frac{1}{2}(\lambda_0-\lambda_{-1})$.

Hence for every $u\in H$ and $t\geq 0$,

$$\| e^{-(\tilde{A}-\lambda^*)t}u\|_H \leq e^{-C_0 t}\|u\|_H \tag{14}$$

Let $t\to u(t)$ be a solution of π on X^α. Then for $t>0$, $u(t)\in D(A)$ and

$t \to u(t)$, $t>0$, is differentiable in X. Hence $t \to u(t)$ is differentiable as a mapping into H and for $t>0$,

$$\dot{u}(t) = -\tilde{A}u(t) + f(u(t)) = -(\tilde{A}-\lambda^*)u(t) + f(u(t)) - \lambda^* u(t) .$$

Fix $t>0$, $h>0$ arbitrarily.
Then by (14)

$$\frac{1}{h}(\|u(t+h)\|_H - \|u(t)\|_H) = \frac{1}{h}(\|e^{-(\tilde{A}-\lambda^*)h}u(t)\|_H - \|u(t)\|_H) +$$

$$+\frac{1}{h}(\|u(t+h)\|_H - \|e^{-(\tilde{A}-\lambda^*)h}u(t)\|_H) \leq \frac{1}{h}(e^{-C_0 h}-1)\|u(t)\|_H +$$

$$+\frac{1}{h}\|(u(t+h)-u(t)) - (e^{-(\tilde{A}-\lambda^*)h}u(t)-u(t))\|_H .$$

Taking $h \to 0^+$ we get

$$\overline{\lim_{h\to 0}} \frac{1}{h}(\|u((t+h)\|_H - \|u(t)\|_H) \leq -C_0\|u(t)\|_H +$$

$$+\|(-\tilde{A}u(t)+f(u(t))) + (\tilde{A}-\lambda^*)u(t)\|_H =$$

$$= -C_0\|u(t)\|_H + \|f(u(t)) - \lambda^* u(t)\|_H .$$

However, our assumptions imply

$$\|f(u(t)) - \lambda^* u(t)\|_H \leq \omega\|u(t)\|_H + \delta$$

where $\omega = \frac{1}{2}(\lambda_{k+1} - \lambda_k) - \varepsilon$ and $\delta > 0$ is a constant.
Therefore

$$\lim_{h\to 0^+} \frac{1}{h}(\|u(t+h)\|_H - \|u(t)\|_H) \leq -(C_0 - \omega)\|u(t)\|_H + \delta.$$

This inequality implies that the function $t \to \|u(t)\|_H$ is decreasing as long as $\|u(t)\|_H$ is large enough. Therefore $t \to \|u(t)\|_H$ must be bounded. Now a standard bootstrapping argument using the Sobolev inequalities (see e.g. Theorem 1.6.1 in Henry [1]) implies that $t \to u(t)$ is bounded in $X = L^p(\Omega)$. An application of Theorem I.4.3 now proves that $t \to u(t)$ is bounded in X^α. Now it is easy to show that π is point-dissipative. Let B be a bounded neighborhood of K_∞ und $u_0 \in X^\alpha$ be arbitrary. Since the solution $u(t) := u_0 \pi t$ is bounded in X^α for $t>0$, it follows from

Theorem I.4.3 that $u(t)$, $t \geq \varepsilon > 0$ lies in a compact set $C \subset X^\alpha$, for arbitrary $\varepsilon > 0$. It follows that $\omega(u_0)$ is a nonempty invariant set for π. Thus, by the definition of K_∞, $\omega(u_0) \subset K_\infty$. But this implies that $u(t) \in B$ for some $t_0 > 0$ and all $t \geq t_0$.

Now our remarks preceding the statement of Theorem 5.7 complete the proof of property (2.1).
The theorem is proved.

Remark
Note that in the extension of Theorem 5.9 to systems ($r>1$) we must, in addition, to an analogue of condition (E*) assume the existence of a function $G: \bar{\Omega} \times \mathbb{R}^r \to \mathbb{R}^r$ such that $f(x,s) = \frac{\partial G}{\partial s}(x,s)$ for all $x \in \bar{\Omega}, s \in \mathbb{R}^r$.

2.6 Estimates at zero and nontrivial solution of elliptic equations

If $f(x,0) \equiv 0$ in Theorem 5.9, then $u=0$ is a solution of equation (1.8), so (3) of Theorem 5.9 is not very informative in this case. Therefore it is useful to have conditions for the existence of nontrivial solutions of equation (1.8). In this section we will present some such conditions. We begin with a theorem which is an improvement for gradient-like systems of Theorem 3.5.

Theorem 6.1

Assume hypotheses (1) and (2) of Theorem 5.5. Furthermore, suppose that $\tilde{A}$ has compact resolvent. Let $\lambda_0 < \lambda_1 < \ldots$ be the eigenvalues of $\tilde{A}$ and let $\lambda_{-1} < \lambda_0$ be arbitrary.
Suppose that $1/2 \leq \alpha < 1$ and let $f: U \to X$ be locally Lipschitzian on a neighborhood U of zero in X^α with $f(0)=0$ and such that the following condition is satisfied:
(F) *There are $\bar{\varepsilon} > 0$ and $l \geq -1$ such that for all $u \in U$*

$$\|f(u) - \mu^* u\|_H \leq \bar{\omega} \|u\|_H$$

where $\mu^* = 1/2(\lambda_l + \lambda_{l+1})$ and $\bar{\omega} = 1/2(\lambda_{l+1} - \lambda_l) - \bar{\varepsilon}$.
Let q^* be the total algebraic multiplicity of all eigenvalues λ of A with $\lambda \leq \lambda_l$.
Suppose also that there is a Fréchet differentiable mapping $G: U \to \mathbb{R}$ such that for all $u \in U$, $h \in X^\alpha$

$$DG(u)(h) = \langle f(u), h \rangle \;.$$

<u>Finally, let</u> π <u>be the local semiflow generated by solutions of</u>

$$\dot{u}+Au = f(u)$$

<u>Then</u>

$$(\pi,\{0\}) \in S(U) \text{ and}$$

$$h(\pi,\{0\}) = \Sigma^{q^*} .$$

<u>Proof</u>:

Theorem 6.1 is dual to a part of Theorem 5.7 and has a dual and even simpler proof. Therefore we shall only sketch the arguments.
Define the homotopy

$$g_\tau(u) = (1-\tau)f(u)+\tau\mu^* u \quad \text{for } \tau \in [0,1],\ u \in U .$$

Let π_τ be the semiflow generated on U by

$$\dot{u}+Au = g_\tau(u) .$$

Using the arguments from the proof of Theorem 5.7 we see that Theorem 6.1 will be proved if we can show that there is a closed bounded set N in X^α, $N \subset U$, such that for every τ, N is an isolating neighborhood of $\{0\}$, relative to π_τ. If this latter statement is not true, then the gradient-like nature of π_τ implies that there is a sequence $\tau_m \in [0,1]$ and a sequence $u_m \in U$, $u_m \neq 0$, such that $u_m \to 0$ in X^α as $m\to\infty$, and u_m is an equilibrium of π_{τ_m} for $m \in \mathbb{N}$.
Proceeding as in the proof of Theorem 5.7 we obtain that for $u \in D$

$$\|-\tilde{A}u+\mu^* u\|_H \geq \overline{C}_0\|u\|_H$$

where $\overline{C}_0-\overline{\omega}>0$.
It follows for every m that

$$\begin{aligned} 0 &= \|-Au_m+(1-\tau_m)f(u_m)+\tau_m\mu^* u_m\|_H = \\ &= \|-\tilde{A}u_m+(\mu^* u_m)+(1-\tau_m)(f(u_m)-\mu^*(u_m))\|_H \geq \\ &\geq \|-\tilde{A}u_m+\mu^* u_m\|_H-(1-\tau_m)\|f(u_m)-\mu^* u_m\|_H \geq \\ &\geq (\overline{C}_0-\overline{\omega})\|u_m\|_H \quad . \end{aligned}$$

It follows that $\|u_m\|_H=0$ i.e. $u_m=0$ a contradiction.
The theorem is proved.

Corollary 6.2

Suppose that all hypotheses of both Theorem 5.7 and Theorem 6.1 are satisfied. Let $k \neq l$.
Then the following properties hold:

(1) *There exists a nontrivial solution u_0 of*

$$Au = f(u)$$

(2) *There exists a nonconstant full bounded solution $t \to u(t)$, $t \in \mathbb{R}$ of*

$$\dot{u} + Au = f(u)$$

such that either $u(t) \to 0$ as $t \to \infty$, or $u(t) \to 0$ as $t \to -\infty$.

Proof:

$k \neq l$ implies that $h(\pi,\{0\}) \neq h(\pi,K_\infty)$. Since $0 \in K_\infty$, it follows that there exists a $v_0 \neq 0$, $v_0 \in K_\infty$. If v_0 is an equilibrium of π then property (1) holds for $u_0 = v_0$.
Otherwise there exists a nonconstant full bounded solution $t \to v(t)$ of π with $v(0) = v_0$. Now by Proposition 5.4 $\omega^*(v)$ and $\omega(v)$ contain equilibria of π, thus completing the proof of (1).
Since $h(\pi,K_\infty) = \Sigma^{m^*}$, (π,K_∞) is irreducible by Theorem I.11.6. Now Theorem I.11.5 implies that there exists a solution $t \to u(t) \in K_\infty$ of π joining $K_1 = \{0\}$ with some set $K_2 \subset K_\infty$, $0 \notin K_2$. This implies property (2) and completes the proof.
Theorem 6.1 and Corollary 6.2 have immediate applications to parabolic and elliptic equations.

Theorem 6.3

Suppose that all hypotheses of Theorem 5.9 hold except (possibly) for condition (E). Instead, assume that f satisfies the following condition (F*):*
(F*) $f(x,0) \equiv 0$ *for all* $x \in \bar{\Omega}$, *and there are* $\bar{\rho}$, $\bar{\varepsilon} > 0$ *and* $l \geq -1$ *such that whenever* $x \in \bar{\Omega}$ *and* $s \in \mathbb{R}$ *with* $0 < |s| \leq \rho$ *then*

$$\lambda_l + \bar{\varepsilon} \leq f(x,s)/s \leq \lambda_{l+1} - \bar{\varepsilon} .$$

Then $(\pi,\{0\})$ *is defined and*

$$(\pi,\{0\}) = \Sigma^{q^*}$$

where q^* *is the total algebraic multiplicity of all eigenvalues* λ *of* A *with* $\lambda \leq \lambda_1$.
If, in addition, f *satisfies condition* (E*) *of Theorem* 5.9 *with* $k \neq 1$, *then the following properties hold:*

(1) *There exists a nontrivial solution* u_0 *of equation* (1.4).
(2) *There exists a nonconstant solution* $t \to u(t)$ *of* (1.7) *such that either* $u(t) \to 0$ *as* $t \to \infty$, *or* $u(t) \to 0$ *as* $t \to -\infty$.

The proof is a simple application of the preceding results. We will now treat the resonance case at zero. We need the following result:

Proposition 6.4

Let N_1, N_2 *be two compact subsets of* $\mathbb{R}^l$, $l \geq 0$, $N_2 \subset N_1$. *Then*

$$\Sigma^\nu \neq [N_1/N_2, [N_2]] \wedge \Sigma^m$$

whenever $\nu, m \geq 0$ *are such that* $\nu < m$ *or* $\nu > m+1$.

Proof:

Write $X = N_1/N_2$ and $x_0 = [N_2]$. Then (X, x_0) is a compact pointed space. For $m \geq 0$, let (S^m, s_0) be the m-dimensional unit sphere with a base point.
Then, by definition, $[N_1/N_2] \wedge \Sigma^m$ is the homotopy type of the pointed space $X \wedge S^m$. Moreover, $S^m \wedge S^k$ is homeomorphic to S^{m+k} (see Proposition 6.2.15 in Maunder [1]).
Moreover, by a restricted associativity property of the smash product, $(X \wedge S^m) \wedge S^k$ is homeomorphic to $X \wedge (S^m \wedge S^k)$ (see Theorem 6.2.23 in Maunder [1]).
It follows that for $m, k \geq 0$, $X \wedge S^{m+k}$ is homeomorphic to $(X \wedge S^m) \wedge S^k$.
Let $(H^q)_{q \in Z}$ be an unreduced cohomology theory. Then by Proposition 7.16 in Switzer [1] $H^q(X, \{x_0\})$ is isomorphic to $H^{q+1}(X \wedge S^1, \{*\})$ for all $q \in Z$. Here, $*$ is the base point of $X \wedge S^1$. (Actually, Proposition 7.16 is a result about homology theory. However, the dual result about cohomology trivially follows by "dualizing" the proof of Proposition 7.16, i.e. by reversing the arrows in all diagrams involved). Consequently, obvious induction implies that

$$H^q(X, \{x_0\}) \cong H^{q+k}(X \wedge S^k, \{*\}) \quad . \tag{1}$$

Here $q \in Z$, $k \geq 0$ and $*$ is base point of $X \wedge S^k$.
Now take $(H^q)_{q \in Z}$ to be the Alexander-Spanier cohomology theory. Let

$\nu, m \geq 0$ be such that $\nu < m$ or $\nu > m+1$. Then (1) implies

$$H^{\nu}(X \wedge S^m, \{*\}) \cong H^{\nu-m}(X, \{x_0\}) \quad . \tag{2}$$

If $\nu < m$, then $H^{\nu-m}(X, \{x_0\}) = 0$, of course.
Moreover, since the dimension of N_1 is ≤ 1, it follows from results in Spanier [1]

$$H^q(X, \{x_0\}) \cong H^q(N_1, N_2) = 0$$

for $q > 1$. (See, in particular, exercise D. 4, Chapter 6 in Spanier [1]).
Hence, in both cases,

$$H^{\nu}(X \wedge S^m, \{*\}) = 0 \quad . \tag{3}$$

In particular, since $H^{\nu}(S^{\nu}, \{s_0\}) \cong \mathbb{R}$, it follows that $X \wedge S^m$ cannot be homotopy equivalent to (S^{ν}, s_0).
This proves the proposition.

Theorem 6.5

Assume all hypotheses of Theorem 5.9 except (possibly) for condition (E*) which is replaced by the following condition

(R) $f(x,0)=0$ for all $x \in \bar{\Omega}$; the partial derivative $\frac{\partial f}{\partial s}(x,s)$ exists for all $|s|$ small and all $x \in \bar{\Omega}$, is continuous in (x,s) and continuous in s uniformly for $x \in \bar{\Omega}$.
Furthermore, there is $\bar{\varepsilon} > 0$ and $k \geq 0, r \geq -1$ such that either

(R1) $r \geq k$ and $\frac{\partial f}{\partial s}(x,0) \leq \lambda_k - \bar{\varepsilon}$ for $x \in \bar{\Omega}$, or else

(R2) $r < k$ and $\frac{\partial f}{\partial s}(x,0) \geq \lambda_k + \bar{\varepsilon}$ for $x \in \bar{\Omega}$.

Under these hypotheses, if $\{0\}$ is an isolated equilibrium of π, then $K=\{0\}$ is an isolated invariant set for π, $h(\pi, \{0\})$ is defined and

$$h(\pi, \{0\}) \neq \textstyle\sum^{m_r} .$$

Here, $m_{-1} := 0$ and for $r \geq 0$ m_r is the total algebraic multiplicity of all eigenvalues λ of A with $\lambda \leq \lambda_r$.

Proof:
If 0 is an isolated equilibrium of π, the gradient-nature of π immediately implies that $K=\{0\}$ is an isolated invariant set. Using hypo-

thesis (R) and the fact $X^{\alpha} \subset C^0(\bar{\Omega})$ with continuous inclusion, it is easily proved that the Nemitski operator $f: X^{\alpha} \to X$ is differentiable in a neighborhood U of 0 and f' is Lipschitz continuous on U.
Let $L = A - f'(0)$. Since A has compact resolvent, L has compact resolvent as well. Therefore the set $\{\lambda \in \sigma(L) \mid \mathrm{re}\, \lambda \leq 0\}$ is a finite set of eigenvalues of L, isolated in $\sigma(L)$. Also, there exists a direct sum decomposition $X = X_1 \oplus X_2 \oplus X_3$, with $\dim(X_1 \oplus X_2) < \infty$, such that X_i is L-invariant for i=1,2,3. Moreover, if $L_i := L|X_i$ then L_i is a sectorial operator on X_i, i=1,2,3 such that $\mathrm{re}\, \sigma(L_1) = 0$, $\mathrm{re}\, \sigma(L_2) < 0$, $\mathrm{re}\, \sigma(L_3) > 0$. (cf. Theorem 1.5.2 in Henry [1]).
Consequently, all hypotheses of Theorem 3.1 are satisfied.
Thus

$$h(\pi, \{0\}) = h(\pi_{\xi}, \{0\}) \wedge \Sigma^m \qquad (4)$$

where $m = \dim X_2$.
We recall that π_{ξ} is the "restriction" of π to a local center manifold.
Let $l = \dim X_1$. Since π_{ξ} is a local semiflow on an open subset of $X_1 \cong \mathbb{R}^l$, we obtain, using an appropriate compact isolating block B for $K_{\xi} = \{0\}$ relative to π_{ξ},

$$h(\pi_{\xi}, \{0\}) = [N_1/N_2, [N_2]] \qquad (5)$$

where $N_1 = B$, $N_2 = B^-$.
Now (5) and Proposition 6.4 imply that

$$h(\pi, \{0\}) \neq \Sigma^{\nu} \qquad (6)$$

whenever $\nu < m$ or $\nu > m+l$.
Now suppose that $\lambda^* \in \tilde{\rho}(A)$. Then $\lambda^* \in \rho(A)$ and there are $(A-\lambda^*)$-invariant subspaces X_+, X_-, $X = X_+ \oplus X_-$, $\dim X_- < \infty$ and such that $\sigma(A_+ - \lambda^* I_+) > 0$ and $\sigma(A_- - \lambda^* I_-) < 0$ where the subscript + or - denotes the restriction of the corresponding operator to X_+ and X_-, respectively.
We claim that for $u \in D(A)$, $u \neq 0$

$$\langle (A-\lambda^*)u, u \rangle \begin{cases} <0 & \text{if } u \in X_- \\ >0 & \text{if } u \in X_+ \end{cases} \qquad (7)$$

and

$$\langle Lu,u\rangle \begin{cases} = 0 & \text{if } u \in X_1 \\ < 0 & \text{if } u \in X_2 \\ > 0 & \text{if } u \in X_3 \end{cases} \qquad (8)$$

Since $\dim X_- < \infty$ and $\dim(X_1 \oplus X_2) < \infty$ and since $A-\lambda^*$ and L are symmetric, the inequalities (7) and (8) follow immediately for $u \in X_- \cup X_1 \cup X_2$.
Now let $\tilde{k}>0$ be such that $\sigma(\tilde{A}-\lambda^*+\tilde{k})>0$.
For $u \in X^\alpha$ define

$$V(u) = \frac{1}{2}(\langle(\tilde{A}-\lambda^*+\tilde{k})^{1/2}u,(\tilde{A}-\lambda^*+\tilde{k})^{1/2}u\rangle-\tilde{k}\langle u,u\rangle).$$

It follows that V is continuous on X^α, Let $u_0 \neq 0$, $u_0 \in D(A)\cap X_+$. If $t \to u(t)$, $t \geq 0$ is the solution through u_0 of

$$\dot{u} = -(A-\lambda^* I)u \qquad (9)$$

then, by well-known results (cf. Theorem 1.5.3, Henry [1]), $u(t) \to 0$ exponentially in X^α, as $t \to \infty$. But since $\frac{dV(u(t))}{dt} = -\|(A-\lambda^* I)u(t)\|_H^2$ for $t>0$, it follows that

$$V(u_0) > V(0) = 0$$

i.e. (7) follows for $u \in D(A)\cap X_+$.
Now define $\Phi: H \to H$ by

$$(\Phi u)(x) = \frac{\partial f}{\partial s}(x,0)\cdot u(x) .$$

Then Φ is a well-defined bounded linear operator, $\Phi(u)=f'(0)(u)$ for $u \in X^\alpha$, $\tilde{L}=\tilde{A}-\Phi$ is self-adjoint and $\tilde{L}(u)=L(u)$ for $u \in X^\alpha$.
Let $\tilde{k}>0$ be such that $\sigma(\tilde{L}+\tilde{k})>0$ and define for $u \in X^\alpha$

$$\tilde{V}(u) = \frac{1}{2}(\langle(\tilde{L}+\tilde{k})^{1/2}u,(\tilde{L}+\tilde{k})^{1/2}u\rangle-\tilde{k}\langle u,u\rangle) .$$

Using $\tilde{V}$ and analogous arguments as above we obtain

$$\tilde{V}(u_0) > \tilde{V}(0) = 0$$

for $u_0 \in D(A)\cap X_3 \setminus \{0\}$, i.e. (8) holds for $u \in D(A)\cap X_3$.
The claim is proved.
Now suppose that (R1) holds. Let $\lambda^* \in \rho(\tilde{A})$ be arbitrary with

$\lambda_k - \bar{\varepsilon} < \lambda^* < \lambda_k$, $\lambda^* > \lambda_{k-1}$ $(\lambda_{-1} := -\infty)$.
Then for $u \in D(A)$

$$\langle Lu,u\rangle \geq \langle (A-\lambda^*)u,u\rangle \ .$$

Hence, (7) and (8) obviously imply that

$$X_+ \cap (X_1 \oplus X_2) = \{0\}.$$

Consequently

$$X_+ \oplus X_1 \oplus X_2 \subset X = X_+ \oplus X_-$$

and therefore $\dim X_1 + \dim X_2 \leq \dim X_-$.
In other words (since $\dim X_- = m_{k-1}$, $m_{-1} := 0$)

$$1 + m \leq m_{k-1} < m_k \leq m_r \ .$$

Now (6) implies the result in case hypothesis (R1) holds. Now suppose (R2) holds.
Choose $\lambda^* \in \rho(A)$, $\lambda_k < \lambda^* < \lambda_k + \bar{\varepsilon}$, $\lambda^* < \lambda_{k+1}$.
Then for $u \in D(A)$

$$\langle Lu,u\rangle \leq \langle (A-\lambda^*)u,u\rangle$$

(7) and (8) imply in this case that

$$X_- \cap (X_1 \oplus X_3) = \{0\}.$$

Hence,

$$X_1 \oplus X_3 \oplus X_- \subset X = X_1 \oplus X_3 \oplus X_2 \ .$$

Therefore $\dim X_- \leq \dim X_2$, i.e. $m_k \leq m$.
Therefore, $m_r < m_k \leq m$ and (6) again implies the result.
The proof is complete.

Corollary 6.6

Assume all hypotheses of Theorem 5.9. Let $f(x,0)=0$ for all $x \in \bar{\Omega}$ and and suppose that $\frac{\partial f}{\partial s}(x,s)$ exists for $|s|$ small and $x \in \bar{\Omega}$ is continuous in (x,s) and continuous in s, uniformly for $x \in \Omega$.

Finally, suppose that the following conditions hold:

(1) *if* $k \geq 0$ *then*

either

(a) $\frac{\partial f}{\partial s}(x,0) \leq \lambda_k - \varepsilon$ *for all* $x \in \overline{\Omega}$

or

(b) $\frac{\partial f}{\partial s}(x,0) \geq \lambda_{k+1} + \varepsilon$ *for all* $x \in \overline{\Omega}$

(2) *if* $k=-1$, *then*

$\frac{\partial f}{\partial s}(x,0) \geq \lambda_0 + \varepsilon$ *for all* $x \in \overline{\Omega}$.

Under these hypotheses there exists a nontrivial solution of (1.4).

Proof:

If 0 is not an isolated equilibrium of π, then there are infinitely many (nontrivial) equilibria, i.e. solutions of (1.4) and we are done. So assume that 0 is an isolated equilibrium of π.
By Theorem 6.5

$$h(\pi,\{0\}) \neq \Sigma^{m_k} \quad .$$

However, Theorem 5.9 implies that

$$h(\pi,K_\infty) = \Sigma^{m_k} \quad .$$

It follows that $0 \in K_\infty$ and $K_\infty \neq \{0\}$. As in the proof of Corollary 6.2 this implies that K_∞ contains a nontrivial equilibrium of π, i.e. a nontrivial solution of (1.4).
The proof is complete.

2.7 Positive heteroclinic orbits of second-order parabolic equations.

In this section we will consider the following equation of type (S_f)

$$\dot{u} + Au = f(u) \qquad (1_f)$$

We make the following assumptions:

(Q_1) Ω is a bounded domain of $\mathbb{R}^q$ with orientable boundary of class $C^{2+\beta}$, $0<\beta<1$. Moreover, for $x \in \Omega$

$$A(x,D) := -\sum_{i,j=1}^{q} a_{ij}(x) D^i D^j - \sum_{i=1}^{q} b_i(x) D^i - c(x) \qquad (2)$$

is a uniformly strongly elliptic differential operator, the functions

$a_{ij}, b_i, c: \overline{\Omega} \to \mathbb{R}$ being Lipschitz continuous. The matrix $(a_{ij}(x))$ is symmetric and $c(x) \leq 0$ for $x \in \overline{\Omega}$.

(Q_2) For $x \in \partial\Omega$

$$B(x,D) = \delta \cdot \frac{\partial}{\partial \mu} + h(x) \tag{3}$$

is a boundary operator with the following properties:

(i) $\mu: \partial\Omega \to \mathbb{R}^q$ is a smoothly varying nontangential direction on Ω pointing outward, and h is of class C^1 and nonnegative.

(ii) either $\delta=0$ and $h \equiv 1$ or else $\delta=1$.

(iii) if $\delta=1$ and $h \equiv 0$ then $c \neq 0$.

(Q_3) For $p>1$, let $A_p: D_p \to L^p(\Omega)$ be the sectorial operator induced by $(A(x,D), B(x,D))$ (see Section 2.1). Then A_2 is self-adjoint. Moreover, $A=A_p$ for some fixed $p>n$, and $f: X^\alpha \to X$, $\alpha \geq \frac{1}{2}$ is the Nemitski operator generated by a function $f: \overline{\Omega} \times \mathbb{R} \to \mathbb{R}$. The function f is Lipschitzian on compact subsets of $\overline{\Omega} \times \mathbb{R}$.

We have the following

Proposition 7.1

Suppose that (Q_1)-(Q_3) *hold.*

Moreover, assume that

(Q_4) $f(x,s) > 0$ *for* $x \in \overline{\Omega}$, $s<0$.

Let $J \subset \mathbb{R}$ *be an open interval and* $u: J \to X^\alpha$ *be a solution of* (1_f) *on* J.

Then the following properties hold:

(1) u *is a classical solution of* (1_f).

(2) *if* $u(t)(x) \geq 0$ *for all* $(x,t) \in \Omega \times J$, *then either* $u \equiv 0$ *or else* $u(t)(x) > 0$ *for all* $(x,t) \in \Omega \times J$.

(3) *if* $J=\mathbb{R}$ *and* u *is bounded, then* $u(t)(x) \geq 0$ *for all* $(x,t) \in \Omega \times \mathbb{R}$.

Proof:

1. Step

u is a classical solution of (1_f): In fact, by Theorem 3.5.2 in Henry [1], the map $t \mapsto \frac{d}{dt} u(t) \in X^\alpha$, $t \in J$ is well-defined and continuous.

Since X^α imbeds continuously into $C^\nu(\overline{\Omega})$, where $0<\nu<2\alpha-\frac{n}{p}$, we conclude that the function $u: \overline{\Omega} \times J \to \mathbb{R}$, $u(x,t) := u(t)(x)$ is well-defined, Hölder continuous with exponent ν in $x \in \overline{\Omega}$ and C^1 in t with the derivative $\frac{d}{dt} u(t) \in C^\nu(\overline{\Omega})$ for $t \in J$.

Moreover, for $t \in J$, the function $x \to f(x,u(x,t))$ is easily seen to be an element of $C^\nu(\overline{\Omega})$. Thus, by (1_f) $Au(t) \in C^\nu(\overline{\Omega})$ for $t \in J$.

By well-known results from the theory of linear elliptic PDEs, there is a unique solution $v \in C^{2+\nu}(\overline{\Omega})$ of the equation

$$Av = g \tag{4}$$

with $g(x)=f(x,u(x,t))-\frac{d}{dt}u(t)(x)$.

Therefore $v=u(t)$ and so $u(.,t) \in C^{2+\nu}(\bar{\Omega})$. All this proves that $u(x,t)=u(t)(x)$ is a classical solution of (1_f).

2. Step

Let $u(x,t)=u(t)(x)\geq 0$ for all $(x,t) \in \Omega\times J$. To prove (2) we may assume w.l.o.g. that u is bounded. Let M be a bound or $|u(x,t)|\,(x,t) \in \bar{\Omega}\times J$. Then, by (Q_3) there is an $L>0$ such that whenever $x \in \bar{\Omega}$, $s,\bar{s} \in [-M,M]$,

$$|f(x,s)-f(x,\bar{s})| \leq L|s-\bar{s}| \tag{5}$$

Let $v(t)(x)=e^{Lt}\,u(t)(x)$. A simple calculation shows that

$$\dot{v}+Av = e^{Lt}(f(u)+Lu) \ . \tag{6}$$

From (5) we get, observing $f(x,0)\geq 0$ (by (Q_4)) and $u(x,t)\geq 0$ that

$$f(x,u(x,t))+Lu(x,t) \geq 0 \tag{7}$$

on $\bar{\Omega}\times J$.

Consequently, $\dot{v}+Av\geq 0$.

Suppose that $u(x_0,t_0)=0$ for some $(x_0,t_0) \in \Omega\times J$. Then, applying the strong maximum principle (pp. 173-174 in Protter and Weinberger [1]) to $w=-v$, we get $v(x,t)=v(t)(x)\equiv 0$ for $x \in \bar{\Omega}$, $t\leq t_0$, $t \in J$. Thus $u(t)\equiv 0$ for all such t. The uniqueness of solutions of (1_f) now implies that $u(t)\equiv 0$ for all $t \in J$, proving (2).

3. Step

Suppose $J=\mathbb{R}$ and u is bounded. Suppose $E=\{(x,t) \in \Omega\times\mathbb{R}\ |u(x,t)<0\}\neq\emptyset$ and let $-M=\inf u$. Then $-M<0$. By assumption (Q_4) and step 1 we get

$$\dot{u}(t)(x)+(Au(t)(x) > 0 \quad \text{for } (x,t) \in E.$$

If there is a point $(x_0,t_0) \in \bar{\Omega}\times\mathbb{R}$ with $u(x_0,t_0)=-M$ then u has its minimum at (x_0,t_0). Therefore we easily obtain that $x_0 \in \partial\Omega$. Now using the strong maximum principle again we obtain

$$\frac{\partial u}{\partial \mu}(x_0,t_0) < 0 \ .$$

Now, by (Q_2),

$$0 = \delta\cdot\frac{\partial u}{\partial\mu}(x_0,t_0)+h(x_0)\cdot u(x_0,t_0)=\delta\cdot\frac{\partial u}{\partial\mu}(x_0,t_0)-M\cdot h(x_0)$$

Hence $\delta=0$ and $h(x_0)=0$, a contradiction to (Q_2).
It follows that there is no $(x_0,t_0)\in\overline{\Omega}\times\mathbb{R}$ with $u(x_0,t_0)=-M$. Thus there is a sequence $(x_n,t_n)\in E$ with $u(x_n,t_n)\to -M$ and $|t_n|\to\infty$. We may assume $x_n\to x_0\in\overline{\Omega}$. Since u is bounded, $\{u(t)\,|\,t\in\mathbb{R}\}$ is relatively compact in X^α, so we may also assume that $u(t_n)\to w$. Since π_f is gradient-like (A_2 being assumed self-adjoint) it follows that w is an equilibrium of π, i.e. a solution of

$$A(w) = f(w).$$

Thus $-M=\min\{w(x)\,|\,x\in\overline{\Omega}\}$.
Applying the strong maximum principle to w we obtain a contradiction. Hence $E=\emptyset$, which proves (3). The proposition is proved.

Theorem 7.2

Assume hypotheses (Q_1)-(Q_4).
Moreover, suppose that there are constants $\gamma_1,\gamma_2\in\mathbb{R}$, $\rho,\varepsilon>0$ such that one of the following properties holds:
either (L_∞):
$\gamma_1+\varepsilon<\lambda_0-\varepsilon$ and
$\gamma_1+\varepsilon\le f(x,s)/s\le\lambda_0-\varepsilon$ for all $x\in\overline{\Omega}$
and all $s\ge\rho$
or (L'_∞)
$\lambda_0+\varepsilon<\gamma_2-\varepsilon$ and
$\lambda_0+\varepsilon\le f(x,s)/s\le\gamma_2-\varepsilon$ for all $x\in\overline{\Omega}$
and all $s\ge\rho$.
Here, λ_0 is the smallest eigenvalue of A.
Under these hypotheses the union K_f^+ of all nonnegative full bounded orbits of π_f is bounded, $h(\pi_f,K_f^+)$ is defined and

$$h(\pi_f,K_f^+) = \begin{cases}\Sigma^0 & \text{if } (L_\infty) \text{ holds}\\ \overline{0} & \text{if } (L'_\infty) \text{ holds.}\end{cases}$$

Proof:

Notice that by results in Amann [1], $\lambda_0>0$ and λ_0 is a simple eigenvalue of A which has an eigenvector u_0 with $u_0(x)>0$ for $x\in\overline{\Omega}$.
Now suppose that (L_∞) holds.
Let $\tilde{f}:\overline{\Omega}\times\mathbb{R}\to\mathbb{R}$ be Lipschitzian on compact sets in $\overline{\Omega}\times\mathbb{R}$ and such that $\tilde{f}(x,s)=f(x,s)$ on $\overline{\Omega}\times[-\rho/2,\infty)$, $\tilde{f}(x,s)<0$ for $s>0$ and $\tilde{f}(x,s)=-\gamma s$ for

$\bar{\Omega}\times(-\infty,-\rho]$, where $\gamma>0$ is such that $-\gamma\leq\gamma_1$.

Let $K^+_{\tilde{f}}$ be the union of all nonnegative full bounded orbits of $\pi_{\tilde{f}}$.

Since (1_f) and $(1_{\tilde{f}})$ define the same equations as long as $u(t)(x)>-\rho/2$ for $x\in\bar{\Omega}$, we obtain that $K^+_{\tilde{f}}=K^+_f$, $h(\pi_f,K^+_f)$ is defined iff $h(\pi_{\tilde{f}},K^+_{\tilde{f}})$ is defined and then the two indices are equal.

An application of Proposition 7.1 shows that $K^+_{\tilde{f}}$ is, in fact, the union of all full bounded orbits of $\pi_{\tilde{f}}$.

Now Theorem 5.9 completes the proof of the theorem in the first case.

Now assume (L'_∞).

Consider the following homotopy:

$$f_\tau(x,s) = (1-\tau)f(x,s)+\tau g(s) \tag{7}$$

for $\tau\in[0,1]$.

Here $g(s)=(\lambda_0+\varepsilon)s+\varepsilon$ for $s\geq 0$ and $g(s)=-s+\varepsilon$ for $s<0$.

Let K_τ be the union of all full bounded orbits of $\pi_\tau:=\pi_{f_\tau}$.

Then, by Lemma 7.3 below there is a closed bounded set $N\subset X^\alpha$ such that $K_\tau\subset \operatorname{Int} N$ for all $\tau\in[0,1]$.

Thus $h(\pi_\tau,K_\tau)$ is defined and

$$h(\pi_0,K_0) = h(\pi_1,K_1) \ . \tag{8}$$

However, $(\pi_0,K_0)=(\pi_f,K^+_f)$ by Proposition 7.1. Therefore the theorem will be proved if we can show that $K_1=\emptyset$.

Indeed, if $K_1\neq\emptyset$, then the fact that π_1 is gradient-like implies that there is an equilibrium u of π_1, i.e. a solution of

$$Au = g(u) \tag{9}$$

By Proposition 7.1, $u\geq 0$ so

$$Au = (\lambda_0+\varepsilon)u+\varepsilon \ . \tag{10}$$

Let $<\ ,\ >$ be the scalar product in $L^2(\Omega)$.

Then

$$<Au,u_0> = (\lambda_0+\varepsilon)<u,u_0>+<\varepsilon,u_0>$$

$$<Au_0,u> = \lambda_0<u_0,u> \ .$$

Since A_2 is self-adjoint and $X^{\alpha}\subset L^2(\Omega)$,

$$(\lambda_0+\varepsilon)<u,u_0>+<\varepsilon,u_0> = \lambda_0<u_0,u> . \tag{11}$$

Hence

$$\varepsilon<u,u_0>+<\varepsilon,u_0> = 0 . \tag{12}$$

Since $<\varepsilon,u_0>=\int_{\Omega}\varepsilon u_0(x)dx>0$, we get a contradiction.
Thus, indeed $K_1=\emptyset$ and the theorem follows.

Lemma 7.3

There exists a constant $L>0$ such that whenever $\tau\in[0,1]$ and $t\to u(t)$ is a full bounded solution of π_τ, then $\|u(t)\|_\alpha\leq L$ for all $t\in\mathbb{R}$.

Proof:

By (L'_∞), there is a constant $M>0$ such that

$$f_\tau(x,s) \geq (\lambda_0+\varepsilon)s-M \tag{13}$$

for all $\tau\in[0,1]$, $x\in\overline{\Omega}$, $s\geq 0$.
If the lemma is not true, then there is a sequence $\tau_n\in[0,1]$ and a sequence $\{u_n\}$ of full bounded solutions of $\pi_n:=\pi_{f_n}$ where $f_n:=f_{\tau_n}$, such that $c_n=\sup_{t\in\mathbb{R}}\|u_n(t)\|_\alpha\to\infty$ as $n\to\infty$ and $c_n-\|u_n(0)\|_\alpha<1$ for all $n\in\mathbb{N}$.
Let $v_n(t)=c_n^{-1}\cdot u_n(t)$. By Proposition 7.1 $v_n(t)\geq 0$ for all $t\in\mathbb{R}$, $n\in\mathbb{N}$. v_n is a full bounded solution of $(1_{\tilde{f}_n})$ where $\tilde{f}_n(x,s)=c_n^{-1}f_n(x,c_ns)$.
Now we obtain from (13)

$$\begin{aligned}<\dot{v}_n(t),u_0> &= -<Av_n(t),u_0>+<\tilde{f}_n(v_n(t)),u_0> = \\ &= -<v_n(t),Au_0>+<\tilde{f}_n(v_n(t)),u_0> \geq \\ &\geq -\lambda_0<v_n(t),u_0>+(\lambda_0+\varepsilon)<v_n(t),u_0>-c_n^{-1}<M,u_0>.\end{aligned} \tag{14}$$

Write $\beta_n(t):=<v_n(t),u_0>$.
Then by (14)

$$\dot{\beta}_n(t) \geq \varepsilon\beta_n(t)-\varepsilon_n \tag{15}$$

where $\varepsilon_n\to 0$ as $n\to\infty$.

This implies that

$$\beta_n(t) \geq \exp(\varepsilon t)(\beta_n(0)-\frac{\varepsilon_n}{\varepsilon})+\frac{\varepsilon_n}{\varepsilon} \quad (16)$$

for $t \geq 0$.
Since $\beta_n(t)$ is bounded, we obtain

$$0 \leq \beta_n(0)-\frac{\varepsilon_n}{\varepsilon} \leq 0 \quad (17)$$

Now, by admissibility, we may assume (taking subsequences if necessary) that $v_n(0) \to v^0$ for some $v^0 \in X^\alpha$. Moreover, by our assumptions

$$\|v^0\|_\alpha = 1 . \quad (18)$$

However, $\langle v^0, u_0 \rangle = \lim_{n\to\infty} \beta_n(0) = 0$ by (17).
Since $v^0 \geq 0$ and $v^0 \neq 0$ by (18) we get a contradiction to the fact that $u_0(x) > 0$ for all $x \in \Omega$.
The lemma is proved.

<u>Theorem 7.4</u>

<u>Assume hypotheses</u> $(Q_1)-(Q_4)$.
<u>Moreover, suppose</u> $f(x,0) \equiv 0$ <u>for</u> $x \in \bar{\Omega}$, <u>and there is an</u> $\tilde{\varepsilon} > 0$ <u>such that either</u>

(L_0): $f(x,s) < \lambda_0 s$ <u>for all</u> $0 < s \leq \tilde{\varepsilon}$, $x \in \bar{\Omega}$,

<u>or</u>

(L_0'): $f(x,s) > \lambda_0 s$ <u>for all</u> $0 < s \leq \tilde{\varepsilon}$, $x \in \bar{\Omega}$.

<u>Under these hypotheses</u>, $\{0\}$ <u>is an isolated</u> π_f<u>-invariant set</u>, $h(\pi_f, \{0\})$ <u>is defined and</u>

$$h(\pi_f,\{0\}) = \begin{cases} \Sigma^0 & \text{if } (L_0) \text{ holds ,} \\ \bar{0} & \text{if } (L_0') \text{ holds.} \end{cases}$$

<u>Proof</u>:

Let $\gamma, \varepsilon > 0$ be arbitrary. Define

$$g(s) = -\gamma s \quad \text{if } (L_0) \text{ holds, and}$$

$$g(s) = \begin{cases} (\lambda_0+\varepsilon)s+\varepsilon & s \geq 0 \\ -s+\varepsilon & s < 0 \end{cases}$$

if (L_0') holds.

Consider the homotopy

$$f_\tau(x,s) = (1-\tau)f(x,s)+\tau g(s) \tag{19}$$

for $\tau \in [0,1]$.
Let N be a closed bounded neighborhood of zero in X^α such that $|u(x)|\leq\tilde{\varepsilon}/2$ for all $u \in N$ and $x \in \bar{\Omega}$ (remember that X^α imbeds continuously into $C^0(\bar{\Omega})$.
Let $\pi_\tau=\pi_{f_\tau}$ and K_τ be the largest π_τ-invariant set in N. We claim that $K_\tau=\{0\}$ for $\tau \in [0,1]$.
Assuming this for the moment we get that $h(\pi_\tau,K_\tau)$ is defined and

$$h(\pi_0,K_0) = h(\pi_1,K_1) \ . \tag{20}$$

Since u=0 is an equilibrium of $\pi_f=\pi_0$, we get that

$$h(\pi_f,\{0\}) = h(\pi_1,K_1) \ . \tag{21}$$

If (L_0) is satisfied, we get $K_1=\{0\}$ and $h(\pi_1,K_1)=\Sigma^0$ by Corollary I.11.2; and if (L_0') holds, then as in the proof of Theorem 7.2, $K_1=\emptyset$ so $h(\pi_1,K_1)=\bar{0}$.
This proves the theorem, except for the claim.
If the claim is not true, then for some $\tau \in [0,1]$, there is a solution $v\neq 0$, $|v(x)|\leq\tilde{\varepsilon}$, $x \in \bar{\Omega}$, of

$$Av = f_\tau(v) \ . \tag{22}$$

By Proposition 7.1, $v(x)>0$ for $x \in \Omega$.
Let u_0 be as in the proof of Theorem 7.2. Evaluating $\langle Av,u_0\rangle$ and $\langle v,Au_0\rangle$ and keeping in mind that A_2 is self-adjoint, we obtain

$$(1-\tau)\langle f(v)-\lambda_0 v,u_0\rangle+\tau\langle g(v)-\lambda_0 v,u_0\rangle = 0 \ . \tag{23}$$

Suppose that (L_0) holds. Then $(f(v)-\lambda_0 v)(x)\leq 0$ and $(g(v)-\lambda_0 v)(x)\leq 0$ for all $x \in \bar{\Omega}$.
If (L_0') holds then $(f(v)-\lambda_0 v)(x)\geq 0$ and $(g(v)-\lambda_0 v)(x)\geq 0$ for all $x \in \bar{\Omega}$.
Hence in both cases, either

$$f(x,v(x)) \equiv \lambda_0 v(x) \quad \text{for all } x \in \bar{\Omega}$$

or else

$$g(v(x)) = \lambda_0 v(x) \quad \text{for all } x \in \bar{\Omega} .$$

However, this contradicts the fact that $v(x)>0$ for $x \in \Omega$ and the definition of g.
The claim is proved.
Theorems 7.2 and 7.4 now imply the following existence result:

Theorem 7.5

Assume hypotheses $(Q_1)-(Q_3)$. *Assume also that* $f(x,0)\equiv 0$ *for* $x \in \bar{\Omega}$. *Moreover, suppose that either* (L_0) *and* (L'_∞) *holds or else that* (L'_0) *and* (L_∞) *holds.*
Under these hypotheses, there exists a positive equilibrium w *of* (1_f) *i.e. a solution of*

$$Aw = f(w) \tag{24}$$

with $w(x)>0$ *for* $x \in \bar{\Omega}$.
Furthermore, there exists a nonconstant full bounded solution $u:\mathbb{R}\to X^\alpha$ *of* (1_f) *with* $u(t)(x)>0$ *for* $x \in \Omega$.
Both w *and* u *are classical solutions of the corresponding equations.*
The proof of Theorem 7.5 is obtained by first redefining $f(x,s)$ for $s<0$, so that $f(x,s)=-\gamma s$ for $s<0$, and then applying our previous results. Details are left to the reader.

2.8 A homotopy index continuation method and periodic solutions of second-order gradient systems

In this section we will study the periodic boundary value problem for second order systems.
To be more precise, let Ω be the bounded interval $[0,T]$, $T>0$, and $f:\mathbb{R}\times\mathbb{R}^m\to\mathbb{R}^m$ be a continuous mapping where $m>0$ is an arbitrary integer. We assume that $f(x+T,u)\equiv f(x,u)$ for $x \in \mathbb{R}$, $u \in \mathbb{R}^m$ and that f is locally Lipschitzian in u, uniformly for $x \in \mathbb{R}$. We are looking for T-periodic solutions $x\to u(x)$ of the following second order system of ordinary differential equations.

$$u''(x)+f(x,u(x)) = 0 \tag{1}$$

Let $X=L^2(\Omega,\mathbb{R}^m)$ and

$$D = \{u \in H^2(\Omega,\mathbb{R}^m)\,|\,u(0)=u(T),\ u'(0)=u'(T)\} .$$

Define

$$A:D\to X \ , \ Au:=-u'' \ . \tag{P}$$

Here, the derivative is understood in the distributional sense. A is self-adjoint and bounded from below. Therefore, A is sectorial (cf. Henry [1]). A has compact resolvent. Moreover, $\sigma(A)$ consists of eigenvalues $\lambda_k := \left(\frac{2\pi}{T}.k\right)^2$, $k=0,1,2,\ldots$.
If X^α, $\alpha\geq 0$, are the fractional power spaces generated by A, then the Nemitski operator $\hat{f}:X^{1/2}\to X$, $\hat{f}(u)(x)=f(x,u(x))$ is well-defined and Lipschitzian on bounded sets in $X^{1/2}$.
Consequently, T-periodic solutions u of (1) can be regarded as equilibria of the local semiflow π on $X^{1/2}$ generated by the solution of

$$\dot{u}+Au = \hat{f}(u) \tag{2}$$

To obtain existence results for solutions of (1) we will first state a basic continuation principle for the homotopy index, which is analogous to and motivated by the continuation principle of the Leray-Schauder degree due to Mawhin.

Consider the following hypotheses:

(Hyp 1) X is a (real) Banach space, $A:D(A)\to X$ is a sectorial operator on X generating the family X^β, $\beta\geq 0$ of fractional power spaces, $0\leq\alpha<1$, and $f:X^\alpha\to X$ is a locally Lipschitzian mapping.

(Hyp 2) The local semiflow π on X^α generated by the solutions of

$$\dot{u}+Au = f(u) \tag{3}$$

is gradient-like with respect to some function $V:X^\alpha\to\mathbb{R}$.

(Hyp 3) A has compact resolvent. Moreover, there is a $\delta>0$ such that $\sigma(A)=\{0\}\cup\sigma'$, re $\sigma'>\delta$.
Let $X_1:=\ker A_1$ and $P:X\to X$ be the projector onto X_1 associated with this spectral decomposition.

(Hyp 4) There is a set $\Gamma\subset X^\alpha$, open in X^α, bounded in X and $f[\Gamma]$ is bounded in X. We write N for the closure of Γ in X^α.

(Hyp 5) For every $\lambda\in(0,1)$, if $u:\mathbb{R}\to X^\alpha$ is a solution of

$$\dot{u} = -Au+\lambda(I-P)f(u)+Pf(u) \tag{S_λ}$$

such that $u[\mathbb{R}]\subset N$, then $u[\mathbb{R}]\subset\Gamma$.

(Hyp 6) Let $\Gamma_1=\{u\in\Gamma \mid Pu=u\}=\Gamma\cap X_1$. Let $\hat{\pi}$ be the local flow on X_1 generated by the ODE

$$\dot{u}_1 = Pf(u_1) \qquad \text{(D)}$$

and $\hat{K}$ be the largest invariant set in $N\cap X_1$. Then $\hat{K}\subset\Gamma_1$ and the homotopy (Conley) index $h(\hat{\pi},\hat{K})$ is non-zero.

Theorem 8.1

If (Hyp 1)-(Hyp 6) are satisfied, then there exists a solution $u_0\in N$ *of*

$$-Au+f(u) = 0 \qquad (4)$$

Proof:

For $\lambda\in[0,1]$ define the following mappings $f_\lambda : X^\alpha\to X$ as

$$f_\lambda(u) = \begin{cases} (1-2\lambda)(I-P)f(u)+Pf(u) & \text{if } 0\leq\lambda\leq 1/2 \\ Pf(Pu+(2-2\lambda)(I-P)u) & \text{if } 1/2\leq\lambda\leq 1 \end{cases}$$

Since for $u\in X^\alpha$, $Pu\in X^\alpha$, $(I-P)u\in X^\alpha$ and since the two definitions agree for $\lambda=1/2$, f_λ is well-defined. Moreover, it is an easy exercise to show that f_λ is locally Lipschitzian for all λ (hence in particular, it makes sense to speak of solutions of (S_λ) and (Hyp 5) above). Finally, if $\lambda_n\to\lambda$ in $[0,1]$ then $f_{\lambda_n}(u)\to f_\lambda(u)$ locally uniformly in X^α.
Consider the equations

$$\dot{u} = -Au+f_\lambda(u) \qquad \text{for } \lambda\in[0,1] \qquad (T_\lambda).$$

Let π_λ be the corresponding local semiflow on X^α generated by the solutions of (T_λ). Theorem I.2.4 implies that whenever $\lambda_n\to\lambda$, then π_{λ_n}, and π does not explode in N. Moreover, (Hyp 3), (Hyp 4) and Theorem I.4.4 imply that for every sequence $\{\lambda_n\}$ in $[0,1]$, N is $\{\pi_{\lambda_n}\}$-admissible.
In particular, (taking the constant sequence $\pi_{\lambda_n}\equiv\pi_\lambda$) we see that N is strongly π_λ-admissible for every $\lambda\in[0,1]$. Let K_λ be the largest invariant set in N relative to π_λ. Since N is strongly π_λ-admissible it follows that K_λ is compact in X^α, and therefore K_λ is closed. If

we can show that $K_\lambda \subset \text{Int } N$ for every $\lambda \in (0,1]$ (i.e. N is an isolating neighborhood of K_λ, relative to π_λ) then the mapping $\Phi(\lambda) = (\pi_\lambda, K_\lambda)$ is a well-defined S-continuous mapping of $\Lambda = (0,1]$ into S (in the sense of Definition I.12.1).

So let us show that $K_\lambda \subset \text{Int } N$ for $\lambda \in \Lambda$. For $0<\lambda<1/2$ this follows from (Hyp 5) since $\Gamma \subset \text{Int } N$. Let $1/2 \leq \lambda \leq 1$ be arbitrary and $u_0 \in K_\lambda$. Then there exists a solution $u: \mathbb{R} \to X^\alpha$ of (T_λ) with $u[\mathbb{R}] \subset N$ and $u(0) = u_0$. Let $u_1(t) := Pu(t)$ $u_2(t) := (I-P)u(t)$. Then $\dot{u}_1(t) := Pf(u_1(t) + (2-2\lambda)\ u_2(t))$ $\dot{u}_2(t) = -A_2 u_2(t)$. Here A_2 is the restriction of A to $X_2 = (I-P)X$. Since A_2 is a sectorial operator on X_2 (see Henry [1], Theorem 1.5.3) with $\sigma(A_2) = \sigma'$, from the fact that re $\sigma' > \delta$, and using simple estimates we get that $u_2(t) = 0$. It follows that $\dot{u}_1(t) = Pf(u_1(t))$ for all $t \in \mathbb{R}$, so $u_1: \mathbb{R} \to X_1$ is a solution of the ODE (D) in (Hyp 6) with $u_1[\mathbb{R}] \subset B$, where $B := N \cap X_1$. By (Hyp 6) $u_1[\mathbb{R}] \subset \Gamma_1$. Thus $u(t) = u_1(t) \in \Gamma \subset \text{Int } N$ for $t \in \mathbb{R}$.

In particular, $u(0) = u_0 \in \text{Int } N$. This proves our claim. Hence, indeed, the map $\Phi: \Lambda \to S$, $\Phi(\lambda) = (\pi_\lambda, K_\lambda)$ is well-defined and S-continuous.

Thus $h(\pi_\lambda, K_\lambda) \equiv \text{const}$ from the continuation invariance of the homotopy index (Theorem I.12.2).

Our argument above proves that for $\lambda \geq 1/2$, $K_\lambda = \hat{K}$ where $\hat{K}$ is as in (Hyp 6). For $\lambda = 1$ the system (T_λ) is uncoupled:

$$(T_1) \text{ is : } \begin{cases} \dot{u}_1 = Pf(u_1) \\ \dot{u}_2 = -A_2 u_2 \end{cases} .$$

Thus we can view π_1 as the product of $\hat{\pi}$ with the semiflow π' on X_2^α generated by $\dot{u}_2 = -Au_2$. As a consequence, we have

$h(\pi_1, K_1) = h(\hat{\pi} \times \pi', \hat{K} \times \{0\}) = h(\hat{\pi}, \hat{K}) \wedge h(\pi', \{0\})$. However, since $\sigma(A_2) > \delta > 0$ it follows from Corollary I.11.2 that $h(\pi', \{0\}) = \Sigma^0$. Consequently, $h(\hat{\pi}, \hat{K}) \wedge \Sigma^0 = h(\hat{\pi}, \hat{K})$ implies that $h(\pi_1, K_1) = h(\hat{\pi}, \hat{K}) \neq \overline{0}$ (by Hyp 6)). Hence for every $\lambda \in (0,1]$, $h(\pi_\lambda, K_\lambda) \neq \overline{0}$ and therefore $K_\lambda \neq \emptyset$. By admissibility, taking a sequence $\lambda_n \to 0$, we see that $K_0 \neq \emptyset$ (although K_0 may have points on ∂N). Choose $v_0 \in K_0$. Since (T_0) is just Eq. (1), there is a solution $u: \mathbb{R} \to X^\alpha$ of equation (1) with $u(0) = v$ and $u[\mathbb{R}] \subset K_0 \subset N$. Since π is gradient-like and K_0 is compact, $\text{Cl}\,u(\mathbb{R})$ contains an equilibrium u_0 of π, i.e. there exists a $u_0 \in N$ with $-Au_0 + f(u_0) = 0$ as claimed. The proof is complete.

<u>Remark</u>:

In (Hyp 3) we may, more generally, assume that $\sigma(A) = \{0\} \cup \sigma' \cup \sigma''$ where re $\sigma' > \delta > 0$ and re $\sigma'' < -\delta < 0$. In this case σ'' necessarily consists of a finite number of eigenvalues with total algebraic multiplicity equal to some $p \geq 0$. In (Hyp 6) we then have to assume that $\Sigma^p \wedge h(\hat{\pi},\hat{K})$ is nonzero, where Σ^p is the homotopy type of a pointed p-sphere. Under these modified assumptions, Theorem 8.1 remains valid with essentially the same proof.

If the set B in (Hyp 6) of Theorem 8.1 is an isolating block for $\hat{\pi}$, then $h(\hat{\pi},\hat{K})=[B/B^-, [B^-]]$.

The next proposition gives a criterion for $[B/B^-,[B^-]]$ to be nonzero.

Proposition 8.2

Let $\tilde{\pi}$ be a local semiflow on a metric space X and let B be an isolating block for π. Assume that B is strongly π-admissible. If the number r_2 of connected components of B^- is finite and greater than the number r_1 of connected components of B, then the homotopy type of $(B/B^-, [B^-])$ is nonzero.

Proof:

Let H^q, $q \geq 0$ be any unreduced cohomology theory (with coefficients in Z). Theorem I.3.7 and Proposition I.10.8 imply that $H^q(B/B^-,[B^-]) \cong H^q(B,B^-)$. There is a long exact sequence

$$\to H^q(B,B^-) \to H^q(B) \xrightarrow{g} H^q(B^-) \xrightarrow{f} H^{q+1}(B,B^-) \to .$$

Now take H^q to be the Alexander-Spanier cohomology theory. Let q=0 and assume that $H^1(B,B^-)=\{0\}$. Then exactness implies that $\operatorname{Im} g=\ker f=H^q(B^-)$, so g is surjective. Since there are only finitely many components in B (resp. B^-), all these components are open in B (resp. B^-). By Corollaries 9 and 6 in Section 5 of Chapter 6 of Spanier [1], we have

$$H^q(B) \cong Z^{r_1} \quad \text{and} \quad H^q(B^-) \cong Z^{r_2} .$$

Since $r_2>r_1$ there is no surjective homomorphism from Z^{r_1} to Z^{r_2}, a contradiction. Hence $H^1(B,B^-)\neq\{0\}$ which proves that the homotopy type of $(B/B^-,[B^-])$ is nonzero.

We shall apply Theorem 8.1 to system (1). Let A be the sectorial operator defined in (P) at the beginning of this section.

Let $P=P_1:X\to X$ be the projector defined as

$Pu \equiv \frac{1}{T}\int_0^T u(x)\,dx$, and write $P_2=I-P$.

Then $X=X_1 \oplus X_2$, $X_1=P_1X$, $X_2=P_2X$ and the sum is orthogonal. If A_i as the restriction of A to X_i, i=1,2, then $A_1=0$ and $D(A_2)=D(A)\cap X_2$ and A_2 is sectorial on X_2 with $\|e^{-A_2t}\|_{X_2}\leq e^{-\delta t}$, t>0 where $\delta>0$ is any number with $\sqrt{\delta}<\frac{2\pi}{T}$. Moreover, for every $\alpha\geq 0$ there is a constant C_α such that $\|A_2^\alpha e^{-A_2t}\|_{X_2}\leq C_\alpha t^{-\alpha}e^{-\delta t}$ (cf. Theorem 1.4.3 in Henry [1]).

We have the following

<u>Lemma 8.3</u>

<u>Let $g:\mathbb{R}\to X_2$ be any bounded locally Hölder continuous mapping. Let M be a bound on g, and $u:\mathbb{R}\to X_2$ be a bounded solution (defined on $\mathbb{R}$!) of the equation</u>

$$\dot{u} = -A_2u+g(t) \quad .$$

<u>Then u is bounded in $X_2^{1/2}$ and</u>

$$\|u(t)\|_{1/2} \leq C_{1/2}M\int_0^\infty s^{-1/2}e^{-\delta s}ds<\infty \quad .$$

<u>Proof</u>:

Let $\alpha=\frac{1}{2}$, $t_0\in\mathbb{R}$ be arbitrary, and L be a bound on u. Then for all n>0 we have

$$A_2^\alpha u(t_0) = A_2^\alpha e^{-A_2n}u(t_0-n) + \int_{t_0-n}^{t_0} A_2^\alpha e^{-A_2(t_0-s)}g(s)ds \ .$$

Thus

$$\|u(t_0)\|_\alpha \leq L\cdot C_\alpha n^{-\alpha}e^{-\delta n}+C_\alpha M\int_{t_0-n}^{t_0}(t_0-s)^{-\alpha}e^{-\delta(t_0-s)}ds$$

$$= LC_\alpha n^{-\alpha}e^{-\delta n}+C_\alpha M\int_0^n s^{-\alpha}e^{-\delta s}ds <$$

$$< LC_\alpha n^{-\alpha}e^{-\delta n}+C_\alpha M\int_0^\infty s^{-\alpha}e^{-\delta s}ds \ .$$

Taking $n\to\infty$ we get

$$\|u(t_0)\|_\alpha \leq C_\alpha M\int_0^\infty s^{-\alpha}e^{-\delta s}ds<\infty .$$

The lemma is proved.

Set $\beta=C'C_{1/2}\int_0^\infty s^{-1/2}e^{-\delta s}$, where C' is a constant such that

$$\|u\|_{C^0([0,T],\mathbb{R}^m)} \leq C'\|u\|_\alpha \quad \text{for } u \in X_2^\alpha .$$

Here we use the fact that X_2^α imbeds in $C^0([0,T],\mathbb{R}^m)$.
We now obtain the following theorem:

Theorem 8.4

Let $f:\mathbb{R}\times\mathbb{R}^m\to\mathbb{R}^m$ be a mapping satisfying the properties listed at the beginning of this section. Moreover, assume that f the gradient of some function $F:\mathbb{R}\times\mathbb{R}^m\to\mathbb{R}$, i.e. $f(x,u):\frac{\partial F}{\partial u}(x,u)$ for all $(x,u)\in\mathbb{R}\times\mathbb{R}^m$. Assume, in addition, that f is bounded on $\mathbb{R}\times\mathbb{R}^m$ by some constant $M>0$. Suppose that there is an open bounded set $G\subset\mathbb{R}^m$ and a class $\mathcal{T}$ of C^1-functions $V:\mathbb{R}^m\to\mathbb{R}$ such that for all $V\in\mathcal{T}$ $G\subset\{v\in\mathbb{R}^m \mid V(v)<0\}$ and for every $u\in\partial G$ there is a $V=V_u\in\mathcal{T}$ such that $V(u)=0$ and whenever $h\in\mathbb{R}^m$ and $\|h\|\leq T^{1/2}M\beta$ then $\langle \operatorname{grad} V(u), f(x,u+h)\rangle\neq 0$ for every $x\in\mathbb{R}$.
Then $B=\operatorname{Cl}G$ is an isolating neighborhood of an invariant set $\tilde{K}\subset B$ relative to the local flow $\tilde{\pi}$ generated by the equation

$$\dot{u} = g(u) \tag{5}$$

where $g(u):=\frac{1}{T}\int_0^T f(x,u)\,dx$.
If the homotopy index $h(\tilde{\pi},\tilde{K})\neq\bar{0}$, then there exists a T-periodic solution $x\to u(x)$ of (1) with $\bar{u}\in\operatorname{Int}B$ and $\|\tilde{u}(x)\|\leq T^{1/2}M\beta$ for all $x\in\mathbb{R}$ where

$$\bar{u} = \frac{1}{T}\int_0^T u(x)\,dx \quad , \quad \tilde{u} = u-\bar{u} .$$

Proof:

Let $f:X^{1/2}\to X$ be the Nemitski operator generated by the function f. We will verify (Hyp 1)-(Hyp 6) of Theorem 8.1. (Hyp 1) and (Hyp 3) follows from our remarks preceding the statement of Lemma 8.3 and those at the beginning of this section. Since f is gradient of F, π is easily seen to be gradient-like with respect to the map $V:X^{1/2}\to\mathbb{R}$ defined as

$$V(u) = 1/2\langle A_1^{1/2}u, A_1^{1/2}u\rangle - k/2\langle u,u\rangle - \int_0^T F(x,u(x))\,dx .$$

Here, $\langle\ ,\ \rangle$ is the scalar product in $X=L^2(0,T,\mathbb{R}^m)$ and $A_1=A+kI$ for some $k>0$. This is verified similarly as in the proof of Theorem 5.5. This proves (Hyp 2).

Now let $G'=\text{int Cl } G=\text{Int } B$. Furthermore let

$$\Gamma=\{u\in X^{1/2}\,|\,P_1u\in G',\|P_2u\|_{1/2}<2MT^{1/2}C_{1/2}\int_0^\infty s^{-1/2}e^{-\delta s}dx\}.$$

Then Γ is open and bounded in X^α. Moreover, $f[\Gamma]$ is bounded. This proves (Hyp 4).

Now let $N=\text{Cl }\Gamma$ and let $u:\mathbb{R}\to X^{1/2}$ be a bounded solution of (S_λ) with $u[\mathbb{R}]\subset N$. Write $u_i=P_iu$, $i=1,2$ for $u\in X$. Since $\|\lambda(I-P)f(u)+Pf(u)\|_X\leq MT^{1/2}$, it follows from Lemma 8.3 that

$$\|u_2(t)\|_{1/2}\leq MT^{1/2}C_{1/2}\int_0^\infty s^{-1/2}e^{-\delta s}ds \tag{6}$$

$$\|u_2(t)\|_{C^0([0,T],\mathbb{R}^m)}\leq MT^{1/2}\beta\ . \tag{7}$$

If $u(t_0)\notin\Gamma$ for some t_0, then by (6) $u_1(t_0)\notin G'$. It follows that $u_1(t_0)\in\partial G$. Then there is a $\tilde V\in\mathcal{T}$ with $\tilde V(u_1(t_0))=0$ and $\langle\text{grad }\tilde V(u_1(t_0)),\ f(x,u_1(t_0)+h)\rangle\neq 0$ for all $x\in\mathbb{R}$ and $h\in\mathbb{R}^m$ with $\|h\|\leq MT^{1/2}\beta$.

In particular, by (7) $\langle\text{grad }\tilde V(u_1(t_0)),\ f(x,u_1(t_0)+u_2(t_0)(x))\rangle\neq 0$ for all $x\in\mathbb{R}$.

Integrating we obtain $\langle\text{grad }\tilde V(u_1(t_0)),\ P_1f(u(t_0))\rangle\neq 0$.

Since $\dot u_1=P_1f(u)$, it follows that $\frac{d}{dt}\tilde V(u_1(t))|_{t=t_0}\neq 0$, implying that for some small $\varepsilon>0$, $\tilde V(u_1(t))>0$ for all $t\in(0,\varepsilon)$ or all $t\in(-\varepsilon,0)$. For all such t, $u(t)\notin\text{Cl }\Gamma=N$, a contradiction.

This proves (Hyp 5). Note that this argument is also valid for $\lambda=0$, hence any full solution of (3) lying in N for $t\in\mathbb{R}$, lies in Γ.

Now $\Gamma_1=\Gamma\cap X_1=G'$ and $N\cap X_1=\text{Cl } G'=\text{Cl } G=B$. Thus $\hat\pi=\tilde\pi$ and $\hat K=\tilde K$. The same argument shows that $\hat K\subset\Gamma_1$ proving that B is an isolating neighborhood of $\tilde K$ relative to $\tilde\pi$, and that (Hyp 6) is satisfied. Now Theorem 8.1 and the remark just made yield a solution $u_0\in\Gamma_1$ of (4). Classical regularity implies that $u_0\in C^2([0,T],\mathbb{R}^m)$ and $u_0(T)=u_0(0)$, $u_0'(T)=u_0'(0)$. Hence extending u_0 periodically for all $x\in\mathbb{R}$ and using the estimate (7) we get the desired result.

The proof is complete.

Remark:

An analogous result can also be proved for the system

$$u_{xx}-\lambda u+f(x,u)\ =\ 0,u\ T\text{-periodic} \tag{8}$$

where $\lambda\in\sigma(A)$.

In fact, let $Au=-u_{xx}+\lambda u$ with $D(A)$ as before. Then $\sigma(A)=\{0\}\cup\sigma'\cup\sigma''$

where re $\sigma'>\delta>0$ and re $\sigma''<-\delta<0$ for some δ.
Now use the remark following the proof of Theorem 8.1.
As a corollary to Theorem 8.4 we obtain

Theorem 8.5

Let f and F satisfy the assumptions of Theorem 8.4, except that f may be unbounded this time.
Suppose that $G\subset\mathbb{R}^m$ is an open bounded set in $\mathbb{R}^m$ and there exists a $\gamma>0$ and a class $\mathcal{T}$ of C^1-functions $V:\mathbb{R}^m\to\mathbb{R}$ with $G\subset\{u|V(u)<0\}$ for all $V\in\mathcal{T}$ and such that whenever $u\in\partial G$, then there is a $V=V_u\in\mathcal{T}$ such that $V(u)=0$ and $\langle \operatorname{grad} V(u), f(x,u+h)\rangle\neq 0$ for all $x\in\mathbb{R}$ and $\|h\|\leq\gamma$. Let $B=\mathrm{Cl}\,G$ and K be the largest invariant set in B with respect to the local flow π generated by the solutions of the ODE

$$\dot{u}(t) = g(u(t)) \tag{9}$$

where

$$g(u) = \frac{1}{T}\int_0^T f(x,u)\,dx .$$

Then $K\subset\mathrm{Int}\,B$. If the homotopy index $h(\pi,K)\neq 0$, then there exists an $\varepsilon_0>0$ such that whenever $0<\varepsilon<\varepsilon_0$, then the equation

$$u_{xx}+\varepsilon f(x,u(x)) = 0 \tag{10}$$

has a T-periodic solution u with

$$\bar{u} = \frac{1}{T}\int_0^T u(x)\,dx\in\mathrm{Int}\,B \ , \ \|u(x)-\bar{u}\| \leq \varepsilon\tilde{M},\ x\in\mathbb{R}\ ,$$

where $\tilde{M}$ is some constant independent of u and ε.

Proof:

First assume that f is bounded. Let $\|f(x,u)\|\leq M$. Choose ε_0 such that $\varepsilon_0 M\,T^{1/2}\beta=\gamma$.
Let $0<\varepsilon\leq\varepsilon_0$, and $\bar{f}(x,u)=\varepsilon f(x,u)$,

$$\bar{F}(x,u) = \varepsilon F(x,u) \ , \ \bar{M} = \varepsilon M .$$

Then for every $u\in\partial G$ there is a $V\in\mathcal{T}$ with $V(u)$ and

$$\langle \operatorname{grad} V(u), f(x,u+h)\rangle \neq 0 \quad \text{for } \|h\|\leq\bar{M}\,T^{1/2}\beta \leq \gamma \tag{11}$$

Let $\tilde{\pi}$ be the local flow generated by

$$\dot{u} = \bar{g}(u) \tag{12}$$

where

$$g(u) := \frac{1}{T}\int_0^T \bar{f}(x,u)dx .$$

Then (11) implies that B is an isolating neighborhood of the largest invariant set $\tilde{K}$ in B relative to $\tilde{\pi}$. Moreover, $t\to u(t)$ is a solution of (9) if and only if $t\to u(\varepsilon t)$ is a solution of (12). In particular, a set $\tilde{B}$ is an isolating block for π if and only if it is so for $\tilde{\pi}$ with the same sets of ingress, resp. egress resp. bounce-off points. It follows that $\tilde{K}=K$ and $h(\tilde{\pi},\tilde{K})=h(\pi,K)\neq\bar{0}$.
Now Theorem 8.4 gives the desired result if f is bounded.
If f is not bounded, take some big ball $B_1=B_{r_1}(0)$ of radius r_1 with $B=\mathrm{Cl}G\subset\mathrm{Int}\ B_1$.
Now let $\varphi:\mathbb{R}^n\to\mathbb{R}$ be a C^∞-function such that

$$\varphi(u) = 1 \quad \text{if } \|u\|\leq 2r_1 ,$$

$$\varphi(0) = 0 \quad \text{if } \|u\|\geq 3r_1 .$$

Let $\tilde{F}(x,u)=\varphi(u)\,.\,F(x,u)$ and $\tilde{f}(x,u)=\frac{\partial F}{\partial u}(x,u)$. Apply the theorem to $\tilde{f}$. Then there are $\tilde{M}>0$, $\varepsilon_0>0$ such that whenever $0<\varepsilon<\varepsilon_0$ then

$$u_{xx}+\varepsilon\tilde{f}(x,u(x)) = 0$$

has a T-periodic solution u with $\bar{u}\in\mathrm{Int}\ B$ and $\|u(x)-\bar{u}\|\leq\varepsilon\tilde{M}$ for all $x\in\mathbb{R}$. Let $\varepsilon_1<\varepsilon_0$ be such that $\varepsilon_1\tilde{M}\leq\frac{1}{2}r_1$. Thus, if $0<\varepsilon<\varepsilon_1$, then for all $x\in\mathbb{R}$

$$\|u(x)\| \leq \|\bar{u}\|+\|u(x)-\bar{u}\| \leq r_1+\frac{1}{2}r_1 < 2r_1 .$$

Thus $\tilde{f}(x,u(x))=f(x,u(x))$, $x\in\mathbb{R}$, so u satisfies the original equation. The proof is complete.

<u>Remark:</u>

There is an analogous result for T-periodic solutions of the system

$$u_{xx}+\lambda u+\varepsilon f(x,u(x)) = 0$$

where

$$\lambda = \left(\frac{2k\pi}{T}\right)^2, k\geq 0$$

(cf. the remark following the proof of Theorem 8.4).
If the Brouwer degree $d(g,G,0)\neq 0$ in Theorem 8.5, then the result is well-known and can be obtained by the coincidence degree method (see Mawhin [1], Gaines-Mawhin [1]). In this case, f needs not be a gradient.
However, as we shall see now, there are situations with f gradient in which $d(g,G,0)=0$, so the coincidence degree method does not work, but the homotopy index $h(\pi,K)\neq 0$ and so Theorem 8.5 can be applied. Also, it will follow from results of section 3.3 in Chapter III, that whenever $d(g,G,0)\neq 0$ in Theorem 8.5 then also $h(\pi,K)\neq 0$. In other words, for gradient systems, whenever the degree method works, then so does the homotopy index, but not conversely.

Proposition 8.6

Define $H:\mathbb{R}^3\to\mathbb{R}$ *by*

$$H(u) = \frac{1}{3}u_1^3 - \frac{1}{2}u_1u_2^2 - \frac{1}{2}u_1u_3^2 - \frac{1}{9}u_1$$

and

$$g(u) = \operatorname{grad} H(u) .$$

Define the functions $V_i:\mathbb{R}^3\to\mathbb{R}$, $i=1,\dots,4$ *as*

$$V_1(u)=u_1-\frac{\sqrt{5}}{3} \qquad V_2(u)=-\left(\frac{u_1}{\sqrt{5}}+1\right)^2+u_2^2+u_3^2$$

$$V_3(u)=-u_1-\frac{\sqrt{5}}{3} \qquad V_4(u)=-\left(\frac{-u_1}{\sqrt{5}}+1\right)^2+u_2^2+u_3^2 .$$

Let $G=\{u\in\mathbb{R}^3 \mid V_2(u)<0$ *for* $i=1,\dots,4\}$.
Then $B=\operatorname{Cl}G$ *is an isolating block for the equation*

$$\dot{u} = g(u) \tag{13}$$

with

$$B^- = \{u\in\partial G \mid V_1(u) = 0 \text{ or } V_2(u) = 0\} .$$

If K *is the largest invariant set in* B *relative to the local semiflow* π *generated by* (13) *then* $h(\pi,K)=$ *homotopy type of* $(B/B^-,[B^-])\neq\bar{0}$.
On the other hand $d(g,G,0)=0$.

Remark: It is easily seen that B has the form

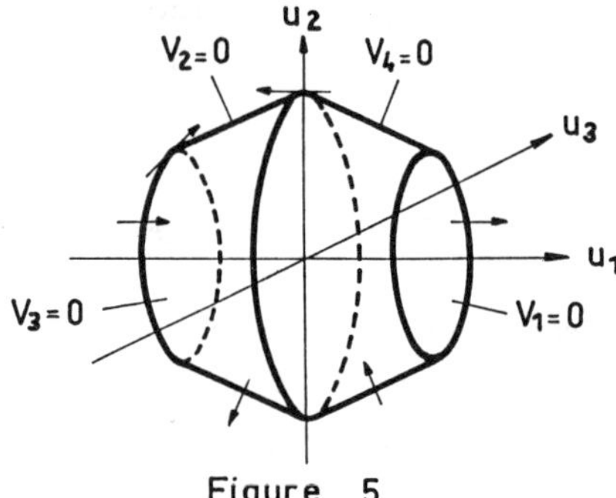

Figure 5

The arrows describe the direction of the flow.

Proof:

$$g_1(u) = u_1^2 - \frac{1}{9} - \frac{1}{2}u_2^2 - \frac{1}{2}u_3^2$$

$$g_2(u) = -u_1u_2$$

$$g_3(u) = -u_1u_3.$$

It follows that $g(u)=0$ if and only if $u_1=\pm\frac{1}{3}, u_2=u_3=0$. Furthermore, $\det Dg(u)=2u_1^3$ if $g(u)=0$.
Moreover, $\left(\pm\frac{1}{3},0,0\right)^T \in G$. Hence $d(g,G,0)$ is defined and equal zero, as claimed.
We will now show that if $u \in \partial G$ and $V_i(u)=0$, then

$$\langle \text{grad } V_i(u), g(u)\rangle > 0 \quad \text{if } i=1 \text{ or } 2$$

and

$$\langle \text{grad } V_i(u), g(u)\rangle < 0 \quad \text{if } i=3 \text{ or } 4 .$$

So let $u \in \partial G$:

1. Case: $V_1(u)=0$.

Then $u_1=\frac{\sqrt{5}}{3}$. Moreover, $V_4(u)\le 0$ implies $u_2^2+u_3^2 \le \left(\frac{-u_1}{\sqrt{5}}+1\right)^2 = \frac{4}{9}$

$$\langle \text{grad } V_1(u), g(u)\rangle = g_1(u) = \frac{4}{9} - \frac{1}{2}\left(u_2^2+u_3^2\right) \ge \frac{2}{9} > 0 .$$

2. Case: $V_2(u)=0$.

Then $u_2^2+u_3^2=\left(\frac{u_1}{\sqrt{5}}+1\right)^2$. Since $V_4(u)\leq 0$, i.e. $u_2^2+u_3^2\leq\left(\frac{-u_1}{\sqrt{5}}+1\right)^2$ we see that $u_1\leq 0$.

Therefore $-\frac{\sqrt{5}}{3}\leq u_1\leq 0$ (as $V_3(u)\leq 0$).

$$<\text{grad } V_2(u),\ g(u)> = -\frac{2}{\sqrt{5}}\left(\frac{u_1}{\sqrt{5}}+1\right)\left(u_1^2-\frac{1}{9}-\frac{1}{2}\left(\frac{u_1}{\sqrt{5}}+1\right)^2\right)-$$

$$-2u_1\left(\frac{u_1}{\sqrt{5}}+1\right)^2 > h(u_1)\left(\frac{u_1}{\sqrt{5}}+1\right)$$

where $h(y)=-\frac{4}{\sqrt{5}}y^2-2y+\frac{2}{9\sqrt{5}}$.

Now $h'(y)<0$ for $y>-\frac{\sqrt{5}}{4}$,

$h'(y)>0$ for $y<-\frac{\sqrt{5}}{4}$.

Therefore

$$h(u_1)\geq \min\left(h\left(-\frac{\sqrt{5}}{3}\right),\ h(0)\right)>0 .$$

Since $\frac{u_1}{\sqrt{5}}+1\geq\frac{2}{3}>0$, we get $<\text{grad } V_2(u),\ g(u)>>0$.

The remaining claims are proved in a similar fashion. Thus, indeed, B is an isolating block for (13) and since $B^-=\{u\in\partial G\,|\,V_1(u)=0\}\dot{\cup}\{u\in\partial G\,|\,V_2(u)=0\}$ has two connected components whereas B is connected, it follows from Proposition 8.2 that the homotopy type of $(B/B^-,[B^-])$ is nonzero, as claimed.

This concludes the proof of the proposition.

<u>Remarks:</u>

If $e:\mathbb{R}\to\mathbb{R}^3$ is any continuous T-periodic function with

$$\int_0^T e(x)\,dx = 0 ,$$

then define $F(x,u)=H(u)+<e(x),u>$ where H is as in the above example. Then F and $f(x,u):=\frac{\partial F}{\partial u}(x,u)$ satisfy all hypotheses of Theorem 8.5

The degree continuation method cannot be used for our choice of G. However, a different choice of G can make the degree method work again. For instance, if $\tilde{G}$ contains just one of the equilibria then $d(g,\tilde{G},0)\neq 0$. However, by a modification of the previous example we can arrange that no choice of $\tilde{G}$ with $d(g,\tilde{G},0)\neq 0$ is possible.

To wit, let $\delta>0$ be small. Let $\beta:\mathbb{R}\to\mathbb{R}$ be a C^∞-function such that $\beta(s)=\beta(-s)$, $s\in\mathbb{R}$,

$$\beta(s) = \begin{cases} 0 & \text{if } |s| \le \frac{1}{3} \\ 1 & \text{if } |s| \ge \frac{1}{3}+\delta \end{cases}$$

$0<\beta(s)<1$ and $\beta'(s)>0$ if $\frac{1}{3}<s<\frac{1}{3}+\delta$.
Define

$$\tilde{H}(u) = \alpha(u_1) - \frac{1}{2}u_1u_2^2 - \frac{1}{2}u_1u_3^2$$

where

$$\alpha(u_1) = \int_0^{u_1} \beta(s)(s^2 - \frac{1}{9})ds .$$

Then for $\tilde{g}=\text{grad}\,\tilde{H}$ we have

$$\tilde{g}_1(u) = \beta(u_1)(u_1^2 - \frac{1}{9}) - \frac{1}{2}u_2^2 - \frac{1}{2}u_3^2$$

$$\tilde{g}_i(u) = g_i(u), \; i=2,3$$

where g is as in the previous example.
If the V_i's are defined as above, and $\delta>0$ is sufficiently small, then it is easy to prove that for $u \in \partial G$

$$\langle \text{grad } V_i(u), \tilde{g}(u) \rangle > 0 \quad \text{if } V_i(u)=0 \text{ and } i=1,2$$

$$\langle \text{grad } V_i(u), \tilde{g}(u) \rangle < 0 \quad \text{if } V_i(u)=0 \text{ and } i=3,4 .$$

Hence $\dot{u}=\tilde{g}(u)$ has the same behavior on the boundary of G as $\dot{u}=g(u)$. In particular, B is again an isolating block with the same set B^-, so the homotopy index is nonzero again.
However, let us show that for any bounded set $U \subset \mathbb{R}^3$ with $\tilde{g}(u) \neq 0$ for $u \in \partial U$ we have $d(\tilde{g},U,0)=0$. In fact, $\tilde{g}(u)=0$ if and only if $|u_1| \le \frac{1}{3}$, $u_2=0=u_3$. Hence the zero set of $\tilde{g}$ is a line segment J. Hence either $U \cap J = \emptyset$ in which case $d(\tilde{g},U,0)=0$ or else $J \subset U$. In the latter case, consider a perturbation $\tilde{g}_\varepsilon$ of $\tilde{g}$, $\tilde{g}=\tilde{g}_\varepsilon(u)-(\varepsilon,0,0)^T$ where $\varepsilon>0$ is a small number. Since $\tilde{g}(u) \neq 0$ on ∂U, we have for all ε small, $\tilde{g}_\varepsilon(u) \neq 0$ on ∂U and $d(\tilde{g}_\varepsilon,U,0)=d(\tilde{g},U,0)$. However, $\tilde{g}_\varepsilon(u)=0$ if and only if

$$\beta(u_1)(u_1^2 - \frac{1}{9}) = \varepsilon \qquad u_2=0=u_3 .$$

Our assumptions imply that for each small ε there are exactly two such solutions

$$u_1 = \pm\gamma(\varepsilon) \qquad \frac{1}{3} < |\gamma(\varepsilon)| < \frac{1}{3}+\delta \ .$$

Computing the Jacobian at the equilibria we get

$$\det D\, \tilde{g}_\varepsilon(u) = u_1^2(\beta(u_1)\cdot 2u_1+\beta'(u_1)\cdot (u_1^2-\frac{1}{9}) \ .$$

If $u_1=+\gamma(\varepsilon)$ then $\beta(u_1)>0$, $\beta'(u_1)>0$, so $\det D\, \tilde{g}_\varepsilon(u)>0$. For $u_1=-\gamma(\varepsilon)$, this determinant is negative. Now for $\varepsilon\to 0$, $\gamma(\varepsilon)\to\frac{1}{3}$, so the two equilibria are both in U for ε small.

This implies $d(\tilde{g}_\varepsilon,U,0)=0$ for ε small, and our claim follows.

The important feature of the above example is the fact that B is simply connected. Due to the Poincaré Bendixson theory, no such example with B simply connected can be given in two dimensions.

On the other hand, there exist simple two-dimensional examples in which B is, say, a ring.

In fact, define $G:\mathbb{R}^2\to\mathbb{R}$ as $G(u)=\|u\|^4-8\|u\|^2$ and let $B=\{u\in\mathbb{R}^2 \mid 1\le\|u\|\le 3\}$. If $g=\operatorname{grad} G$, then B is easily seen to be an isolating block for $\dot{u}=g(u)$ with $B^-=\partial B$.

Thus $h(K)\neq\overline{0}$. On the other hand, it is not hard to prove (by suitably perturbing g), that $d(g,U,0)=0$, where $U=\operatorname{Int} B$.

Chapter III

Selected topics

We begin this chapter by the study of the inner structure of an invariant set.
We define repeller-attractor pairs and Morse decompositions and prove their basic properties (Section 3.1).
Then we show that given an (admissible) isolated invariant set K and a repeller-attractor pair (A^*,A) in K, there exists a block pair for (A^*,A), i.e. a block B for K which is the union of a block B_1 for A^* and a block B_2 for A with $B_1 \cap B_2 = B_1^- \cap B_2^+$. We define index triples and show that block pairs generate special index triples satisfying the cofibration property (Section 3.2).
Using index triples, we prove, in Section 3.3, an important Morse equation for a Morse decomposition $(M_1,\dots,M_n)$ of K, relating the Betti numbers of $h(M_i)$ to the Betti numbers of $h(K)$. The cofibration property of our special index triples enables us to establish the Morse equation for arbitrary homology or cohomology theory, which even in the compact case is an improvement of Conley's original theory, in which only the Alexander-Spanier cohomology theory can be used. The main application of the Morse equation is, of course, to multiplicity results in variational problems. We illustrate this in a simple case (Proposition 3.7). We do not give any further multiplicity results, reserving them for a subsequent volume.
We end the section by applying the Morse equation to establish a formula relating the Conley index to the Brouwer degree for gradient systems.
Next we examine the relation between admissibility and the Palais-Smale condition for ODEs in Banach spaces. We also show that the so-called critical groups of a critical point p are just the homology or cohomology groups of the homotopy index of $\{p\}$ (Section 3.4).
In the final Section 3.5 we establish the homotopy invariance property of the categorial Morse index along paths.

3.1 Repeller-attractor pairs and Morse-decompositions

In this section we will continue our study of the inner structure of an invariant set, begun in Section 1.11. We define repeller-attractor pairs and Morse decompositions and prove their basic properties.
As usual, X is a metric space and π is a (fixed) local semiflow on X.
Let $Y \subset X$ be such that $\omega_x = \infty$ for all $x \in Y$.
Define $\omega(Y) := \bigcap_{t \geq 0} Cl(Y\pi[t,\infty))$.
$\omega(Y)$ is called the *collective ω- limit set of elements of* Y.

Proposition 1.1

1. *For all $y \in X$, y is in $\omega(Y)$ if and only if that are sequences $x_n \in Y$ and $t_n \geq 0$, such that $t_n \to \infty$ and $x_n \pi t_n \to y$ as $n \to \infty$.*
2. *$\omega(Y)$ is closed.*
3. *If $Cl(Y\pi[0,\infty))$ is compact, then $\omega(Y)$ is compact and invariant.*

Proof:

1. Suppose $y \in \omega(Y)$ and τ_n be a sequence with $\tau_n \to \infty$. Then for every n there is a $y_n \in Y\pi[\tau_n,\infty)$, $d(y,y_n) < 1/n$.
y_n has the form $y = x_n \pi t_n$ where $x_n \in Y$ and $t_n \geq \tau_n$. Since $\tau_n \to \infty$ it follows that $t_n \to \infty$ and this proves the first the only part of 1. The converse is trivial.

2. Trivial.

3. The proof of 3. can be easily accomplished using, say, arguments from the proof of Theorem I.4.5.

Definition 1.2

Let S be a compact and (not necessarily isolated) invariant subset of X.
A subset $A \subset S$ is called an *attractor* (*in* S) if there is a neighborhood U of A in X such that $\omega(U \cap S) = A$. If A is an attractor, then the set

$$A^* := \{x \in S \mid \omega(x) \cap A = \emptyset\}$$

is called the *repeller dual to* A (*relative to* S), and the pair (A^*, A) is called a *repeller-attractor pair in* S.

Remark:

Since S is compact, we have $\omega_x = \infty$ for every $x \in S$, so $\omega(U \cap S)$ is well-defined.

Proposition 1.3

Let (A^*,A) be a repeller-attractor pair in S. Then the following properties hold:

a) If V is open in X with $V\supset A$ then there is a $t_0=t_0(V)$ such that $x\pi t\in V$ for all $x\in U\cap S$ and $t\geq t_0$.

b) If B is closed and disjoint from A, then for every $\varepsilon>0$ there is a $t_0=t_0(\varepsilon)$ such that $d(x,A^*)<\varepsilon$ whenever $x\in S$ and $t\geq t_0$ is such that $x\pi t\in B$.

Proof:

If a) is not true then there is an open set $V\supset A$ and sequences $x_n\in U\cap S$, $t_n\to\infty$ such that $x_n\pi t_n\notin V$ for all n. The compactness of S implies that, w.l.o.g., $x_n\pi t_n$ converges to some $y\in S\setminus V$. By Proposition 1.1, $y\in\omega(U\cap S)=A$, a contradiction.

If b) is not true, then there is a closed set B, $B\cap A=\emptyset$, an $\varepsilon>0$, and sequences $x_n\in S$, $t_n\to\infty$, $x_n\pi t_n\in B$ and $d(x_n,A^*)\geq\varepsilon$. Without loss of generality, $x_n\to x\in S$, and so $d(x,A^*)\geq\varepsilon$. Hence $x\notin A^*$ and so $\omega(x)\cap A\neq\emptyset$. Thus $x\pi t_1\in U\cap S$ for some $t_1\geq 0$. By continuity, $x_n\pi t_1\in U\cap S$ for all n large enough. Taking $V:=X\setminus B$ and $t_0:=t_0(V)$ as in a) we obtain $x_n\pi t\in V$ for all $t\geq t_0+t_1$, a contradiction to the assumption that $t_n\to\infty$.

Now the following basic result holds:

Theorem 1.4

Let (A^*,A) be a repeller-attractor pair in S. Then

(i) A and A^* are disjoint, compact and invariant.

(ii) If $\sigma:\mathbb{R}\to S$ is a full solution through $y\in S$, then the following properties hold: (a) if $y\in A^*$ or if $\omega(y)\cap A^*\neq\emptyset$, then $\sigma[\mathbb{R}]\subset A^*$. (b) if $\omega^*(\sigma)\cap A\neq\emptyset$ then $\sigma[\mathbb{R}]\subset A$. (c) if $y\notin A^*\cup A$, then $\omega^*(\sigma)\subset A^*$ and $\omega(y)\subset A$.

Proof:

(i): Clearly $A\cap A^*=\emptyset$, and it follows from Definition 1.2, Proposition 1.1, and from the compactness of S that A is compact and invariant. Obviously A^* too is invariant.

Let $x_n\in A^*$ and $x_n\to x\in S$. Then $\omega(x)\cap A=\emptyset$ or else $x\pi t\in U$ and so $x_n\pi t\in U$ for some $t>0$ and all n sufficiently large. Consequently $\omega(x_n)=\omega(x_n\pi t)\subset\omega(U\cap S)=A$, a contradiction which proves that $\omega(x)\cap A=\emptyset$ i.e. $x\in A^*$. Thus A^* is closed and consequently compact. This proves (i).

Now let $\sigma:\mathbb{R}\to S$ be a full solution of π through y. We will prove (ii) part a).

Suppose that $y \in A^*$ or $\omega(y) \cap A^* \neq \emptyset$ and pick a closed neighborhood B of A^* with $B \cap A = \emptyset$. Then there is a sequence $t_n \to \infty$ such that $\sigma(t_n) \in B$. Let $t \in \mathbb{R}$ and let $\varepsilon > 0$, then $t_n - t \geq t_0(\varepsilon)$ for n large, where $t_0(\varepsilon)$ is as in b) of Proposition 1.3. Since $\sigma(t)\pi(t_n - t) = \sigma(t_n) \in B$ we conclude that $d(\sigma(t), A^*) < \varepsilon$. This holds for every $\varepsilon > 0$, and so $\sigma(t) \in A^*$. This proves a). To prove b) assume $\omega^*(\sigma) \cap A \neq \emptyset$, so that there is a sequence $t_n \to \infty$ with $\sigma(-t_n) \in U \cap S$, U being as in Definition 1.2. If $t \in \mathbb{R}$, then for n large, $t_n + t \geq 0$, and so $\sigma(-t_n)\pi(t_n + t) = \sigma(t)$. Therefore $\sigma(t) \in \omega(U \cap S)$, i.e. $\sigma(t) \in A$. Finally, to prove (c) assume that $y \notin A^* \cup A$, and let $x \in \omega^*(\sigma)$, so that $\sigma(-t_n) \to x$ for some sequence $t_n \to \infty$. It then follows from Proposition 1.3, b) applied to $B = \{y\}$, that for $\varepsilon > 0$, $d(\sigma(-t_n), A^*) < \varepsilon$ if n is sufficiently large, and thus $x \in A^*$. If on the other hand $x \in \omega(y)$, then $\sigma(t_n) \to x$ for a sequence $t_n \to \infty$. We claim that for some t_{n_0} we have $\sigma(t_{n_0}) \in U$ with U as in Definition 1.2. In fact otherwise $\sigma(t_n) \in S \setminus U$ for all n, and we conclude by choosing $B = S \setminus U$, that $y \in A^*$ contradicting the assumption on y. From $\sigma(t_{n_0}) \in U \cap S$ and $\omega(U \cap S) = A$ we conclude $x \in A$. This concludes the proof of Theorem 1.4.

Definition 1.5

Let S be a compact and (not necessarily isolated) invariant subset of X. An ordered collection $(M_1, \ldots, M_n)$ of subsets $M_j \subset S$ is called a Morse decomposition of S, if there exists an increasing sequence

$$\emptyset = A_0 \subset A_1 \subset A_2 \subset \ldots \subset A_n = S$$

of attractors (in S) such that

$$M_j = A_j \cap A_{j-1}^*, \quad 1 \leq j \leq n .$$

For example, if A is an attractor in S, then (A, A^*) is a Morse decomposition of S. In fact, set $A_0 = \emptyset$, $A_1 = A$ and $A_2 = S$, then $M_1 = A$ and $M_2 = A^*$. The following simple results generalize the remarks made at the beginning of this section.

Proposition 1.6

Suppose that π is gradient-like with respect to a function $V: X \to \mathbb{R}$. Let S be a compact invariant subset of X containing only a finite number of equilibria a_i, $i = 1, \ldots, n$. Order the equilibria so that $V(a_i) \leq V(a_{i+1})$ for $i = 1, \ldots, n-1$.

Then $(M_1,\ldots,M_n)$ *where* $M_i=\{a_i\}$, $i=1,\ldots,n$, *is a Morse decomposition of* S.

The proof of Proposition 1.5 will be a simple consequence of Theorem 1.8 below.

The basic properties of Morse decomposition are contained in the following

Theorem 1.7

Let $(M_1,\ldots,M_n)$ *be a Morse decomposition of* S *and let* $\emptyset=A_0\subset A_1\subset\ldots\subset A_n=S$ *be an associated sequence of attractors. Then the following properties are satisfied:*

(i) *The sets* M_j *are pairwise disjoint.*

(ii) *If* $y\in S$ *and* $\sigma:\mathbb{R}\to S$ *is any full solution through* y, *then either* $\sigma[\mathbb{R}]\subset M_j$ *for some* j *or else there are indices* $i<j$ *such that* $\omega^*(\sigma)\subset M_j$ *and* $\omega(y)\subset M_i$.

(iii) *The attractors are uniquely determined by* $(M_1,\ldots,M_n)$, *namely* $A_k=\{y\in S\mid$ *there is a full solution* $\sigma:\mathbb{R}\to S$ *through* y *with* $\omega^*(\sigma)\subset M_1\cup\ldots\cup M_k\}$ *for* $1\le k\le n$.

(iv) *For every* $i=1,\ldots,n$, (M_i,A_{i-1}) *is a repeller-attractor pair in* A_i.

(v) *If* S *is an isolated invariant set which has a strongly* π*-admissible isolating neighborhood, then the same is true for the sets* M_i, $i=1,\ldots,n$, *and* A_i, $i=0,\ldots,n$.

Proof:

(i) If $i<j$, then $M_i\cap M_j=A_i\cap A^*_{i-1}\cap A_j\cap A^*_{j-1}=A_i\cap A^*_{j-1}\subset A_{j-1}\cap A^*_{j-1}=\emptyset$, by Theorem 1.4, hence the sets M_j are pairwise disjoint.

(ii) Let $y\in S$ and let $\sigma:\mathbb{R}\to S$ be any full solution through y. Since $A_n=S$ and $A^*_0=S$ there is a smallest integer i such that $\omega(y)\subset A_i$, and there is a largest integer j such that $\omega^*(\sigma)\subset A^*_j$. Clearly $i>0$ and $j<n$. Now $\omega(y)\not\subset A_{i-1}$, so $y\in A^*_{i-1}$ by Theorem 1.4.
The same theorem implies that $\sigma[\mathbb{R}]\subset A^*_{i-1}$ and $\omega(y)\subset A^*_{i-1}$. On the other hand, $\omega^*(\sigma)\not\subset A^*_{j+1}$ and we claim that $\sigma[\mathbb{R}]\subset A_{j+1}$. In fact, otherwise $\sigma(t)\notin A_{j+1}$ for some $t\in\mathbb{R}$. If now $\sigma(t)\notin A^*_{j+1}$ then from Theorem 1.4 we conclude that $\omega^*(\sigma)\subset A^*_{j+1}$ a contradiction. Thus $\sigma(t)\in A^*_{j+1}$ and so $\omega(\sigma)\subset A^*_{j+1}$, again a contradiction. Hence, indeed, $\sigma[\mathbb{R}]\subset A_{j+1}$.
Now $j\ge i-1$, since otherwise $j+1\le i-1$ and thus $A_{j+1}\subset A_{i-1}$, which implies $\sigma(\mathbb{R})\subset A_{i-1}\cap A^*_{i-1}=\emptyset$. If $j=i-1$, then $\sigma[\mathbb{R}]\subset A^*_{i-1}\cap A_i=M_i$. If $j>i-1$, then $\omega(y)\subset A^*_{i-1}\cap A_i=M_i$ and $\omega^*(\sigma)\subset A^*_j\cap A_{j+1}=M_{j+1}$. This proves (ii).

(iii) Let $y\in A_k$. Since A_k is invariant, there is a full solution

$\sigma:\mathbb{R}\to A_k$ through y and so $\omega^*(\sigma)\subset A_k$. Let $i\leq k$ be the smallest integer such that $\omega^*(\sigma)\subset A_i$. Then $i>0$ and $\omega^*(\sigma)\not\subset A_{i-1}$ and hence $\omega^*(\sigma)\subset A^*_{i-1}$ by Theorem 1.4. Therefore $\omega^*(\sigma)\subset A_i\cap A^*_{i-1}=M_i\subset(M_1\cup\ldots\cup M_k)$. Conversely, suppose that there is a solution $\sigma:\mathbb{R}\to S$ through y such that $\omega^*(\sigma)\subset M_1\cup\ldots\cup M_k$. Then $\omega^*(\sigma)\subset M_j$ for some $j\leq k$, hence $\omega^*(\sigma)\subset A_j\subset A_k$ and so $\sigma[\mathbb{R}]\subset A_k$ by Theorem 1.4. This proves (iii).
(iv) Follows trivially from Theorem 1.4 part (i) and Definition 1.2.
(v) Let N be an isolating neighborhood of S. We shall prove that each M_j, $j=1,\ldots,n$ is an isolated invariant set. Since $A_j\cap A^*_j=\emptyset$ there is an $\varepsilon>0$ such that $d(x,y)\geq\varepsilon$ for $x\in A_j$ and $y\in A^*_j$, $1\leq j\leq n$. Pick some j, $1\leq j\leq n$, and choose $0<\delta\leq\varepsilon/2$ so that $\hat{N}:=\{x\,|\,d(x,M_j)\leq\delta\}\subset\operatorname{Int} N$. Clearly $M_j\subset\operatorname{Int}\hat{N}$. If K is the largest invariant set contained in $\hat{N}$, then $M_j\subset K\subset S$. Suppose $K\setminus M_j\neq\emptyset$. Let $y\in K\setminus M_j$ and $\sigma:\mathbb{R}\to\hat{N}$ be a full solution through y. Since $y\notin M_j$ we have $y\notin A_j$ or $y\notin A^*_{j-1}$. If $y\notin A_j$ then, by Theorem 1.4, $\omega^*(\sigma)\subset A^*_j$ and so $A^*_j\cap\hat{N}\neq\emptyset$. Therefore there are $x\in A^*_j$ and $x_0\in M_j=A_j\cap A^*_{j-1}$ with $d(x,x_0)\leq\delta$ contradicting $d(x,x_0)\geq\varepsilon$. If on the other hand $y\notin A^*_{j-1}$, then by Theorem 1.4 we have $\omega(y)\in A_{j-1}$ and so $A_{j-1}\cap\hat{N}\neq\emptyset$. It follows that there are $x\in A_{j-1}$ and $x_0\in M_j=A_j\cap A^*_{j-1}$ with $d(x,x_0)\leq\delta$, again a contradiction. We conclude that $\hat{N}$ is an isolating neighborhood of M_j, which, in addition, is strongly π-admissible if $\hat{N}$ is strongly π-admissible.
We will show that A_j, $j=0,\ldots,n$ is isolated. In fact, since $A_j\cap A^*_j=\emptyset$ we can choose a closed neighborhood $\tilde{N}$ of A_j such that $\tilde{N}\subset N$ and $\tilde{N}\cap A^*_j=\emptyset$. Let $\tilde{K}$ be the largest invariant set contained in $\tilde{N}$, then $A_j\subset\tilde{K}\subset S$. We claim that $A_j=\tilde{K}$. In fact, suppose $\tilde{K}\setminus A_j\neq\emptyset$, pick $\tilde{y}\in\tilde{K}\setminus A_j$ and let $\tilde{\sigma}:\mathbb{R}\to\tilde{K}$ be a full solution through $\tilde{y}$. Since $\tilde{y}\notin A_j\cup A^*_j$ we conclude that $\omega^*(\sigma)\subset A^*_j$ and hence $A^*_j\cap\tilde{N}\neq\emptyset$, in contradiction to the choice of $\tilde{N}$.
The theorem is proved.
Properties (i) and (ii) of Theorem 1.7 uniquely characterize a Morse decomposition.
More precisely, we have

Theorem 1.8

Let S be as in Theorem 1.7 and let $(M_1,\ldots,M_n)$ be an ordered collection of pairwise disjoint compact and invariant subsets of S. Suppose that for every $y\in S$ and every full solution $\sigma:\mathbb{R}\to S$ through y either $\sigma[\mathbb{R}]\subset M_j$ for some j or else there are indices $i<j$ such that $\omega^*(\sigma)\subset M_j$ and $\omega(\sigma)\subset M_i$. The $(M_1,\ldots,M_n)$ is a Morse decomposition of S.

Proof:
Set $A_0:=\emptyset$ and for $1\leq k\leq n$ define $A_k:=\{y\in S\,|$ there is a full solution

$\sigma:\mathbb{R}\to S$ through y satisfying $\omega^*(\sigma)\subset(M_1\cup\ldots\cup M_k)\}$. We shall show that $A_0\subset A_1\subset A_2\subset\ldots\subset A_n=S$ is a sequence of attractors in S such that $A_i\cap A^*_{i-1}=M_i$, thus proving the statement.

1. Step:

The sets A_k, $1\le k\le n$ are closed: Since by definition $A_n=S$, the set A_n is closed. We now proceed inductively and assume A_{k+1} to be closed for some $1\le k\le n-1$. Let $y_m\in A_k$ with $y_m\to y\in S$. Then $y\in A_{k+1}$, since $A_k\subset A_{k+1}$ and A_{k+1} is closed. There are full solutions $\sigma_m:\mathbb{R}\to S$ with $\sigma_m(0)=y_m$ and $\omega^*(\sigma_m)\subset M_1\cup\ldots\cup M_k$. Using the compactness of S one finds a subsequence, again denoted by $\{\sigma_m\}$, which converges pointwise to a solution $\sigma:\mathbb{R}\to S$ through y. We claim that $\omega^*(\sigma)\subset(M_1\cup\ldots\cup M_k)$. Indeed, since $\sigma_m[\mathbb{R}]\subset A_k\subset A_{k+1}$ and A_{k+1} is closed, it follows that $\sigma[\mathbb{R}]\subset A_{k+1}$ and so $\omega^*(\sigma)\subset A_{k+1}$. Observe that $M_j\cap A_{k+1}=\emptyset$ for $j>k+1$ since M_j is invariant. On the other hand, $\omega^*(\sigma)\subset M_j$ for some j by our assumptions and therefore $\omega^*(\sigma)\subset M_1\cup\ldots\cup M_k\cup M_{k+1}$. Consequently, either $\omega^*(\sigma)\subset M_1\cup\ldots\cup M_k$ in which case we done, or else $\omega^*(\sigma)\subset M_{k+1}$. In the latter case, let $V\supset M_{k+1}$ be an open neighborhood of M_{k+1} such that $\mathrm{Cl}\,V\cap M_j=\emptyset$ for $j\ne k+1$. There is a sequence $t_\nu\to\infty$ and a $z\in M_{k+1}$ such that $\sigma(-t_\nu)\in V$ and $d(\sigma(-t_\nu),z)\le\nu^{-1}$ for all $\nu\ge 1$. Therefore, for every ν there is a $m_\nu\ge\nu$ such that $\sigma_{m_\nu}(-t_\nu)\in V$ and $d(\sigma_{m_\nu}(-t_\nu),z)\le 2\cdot\nu^{-1}$. Since $\omega^*(\sigma_m)\cup\omega(\sigma_m)\subset(M_1\cup\ldots\cup M_k)$ for every m, there are $\tau_\nu<t_\nu<s_\nu$ such that $\sigma_{m_\nu}(-s_\nu)$ and $\sigma_{m_\nu}(-\tau_\nu)\in\partial V$ and $\sigma_{m_\nu}(-t)\in\mathrm{Cl}\,V$ for $\tau_\nu\le t\le s_\nu$. The invariance of M_{k+1} now implies that $t_\nu-\tau_\nu\to\infty$. Let $x_\nu:=\sigma_{m_\nu}(-s_\nu)$, then $x_\nu\in S$ and since S is compact we may assume $x_\nu\to x\in\partial V$. It then follows that $x\pi t\in\mathrm{Cl}\,V$ for all $t\ge 0$ and so $\omega(x)\in\mathrm{Cl}\,V$ which implies by our hypotheses that $\omega(x)\subset M_{k+1}$. Since A_{k+1} is closed we have $x\in A_{k+1}$ and so there is a full solution $\tilde\sigma:\mathbb{R}\to S$ through x with $\omega^*(\tilde\sigma)\subset M_1\cup\ldots\cup M_{k+1}$. The ordering of the sets M_j implies that $\tilde\sigma[\mathbb{R}]\subset M_{k+1}$ and so $x\in M_{k+1}$. This contradicts $x\in\partial V$ as $M_{k+1}\cap\partial V=\emptyset$. Step 1 is proved.

2. Step:

A_k is an attractor in S, $1\le k\le n$: This, in fact, is true for k=n. We proceed by induction and assume A_{k+1} to be an attractor in S for some $k\le n-1$. Choose a neighborhood $U_{k+1}\supset A_{k+1}$ of A_{k+1} such that $\omega(U_{k+1}\cap S)=A_{k+1}$. Since A_k is closed, $M_{k+1}\cup A_k\subset A_{k+1}$ and $M_{k+1}\cap A_k=\emptyset$, we can choose a neighborhood U_k of A_k and a neighborhood V of M_{k+1}, both contained in U_{k+1} such that $\mathrm{Cl}\,U_k\cap\mathrm{Cl}\,V=\emptyset$. Since A_k is invariant and contained in U_k we have $A_k\subset\omega(U_k\cap S)$. Suppose $\omega(U_k\cap S)\setminus A_k\ne\emptyset$, and choose $y\in\omega(U_k\cap S)\setminus A_k$. Then there are sequences $x_n\in U_k\cap S$ and $t_n\to\infty$ such that

$x_n \pi t_n \to y$. We may assume that $x_n \pi (t_n + t) \to \sigma(t)$ for every $t \in \mathbb{R}$ where σ is a solution $\sigma: \mathbb{R} \to S$ through y. Now $\omega(U_k \cap S) \subset \omega(U_{k+1} \cap S) = A_{k+1}$ implies $\sigma[\mathbb{R}] \subset A_{k+1}$, whence, by step 1, $\omega^*(\sigma) \subset A_{k+1}$ and so $\omega^*(\sigma) \subset (M_1 \cup \ldots \cup M_{k+1})$. But $y \notin A_k$ and so $\omega^*(\sigma) \subset M_{k+1}$. There is a sequence $\rho_\nu \to \infty$ and a $z \in M_{k+1}$ such that $\sigma(-\rho_\nu) \in V$ and $d(\sigma(-\rho_\nu), z) \leq \nu^{-1}$ for every ν. Therefore, for every ν there is $n_\nu \geq \nu$ such that $t_{n_\nu} > \rho_\nu$, $x_{n_\nu} \pi (t_{n_\nu} - \rho_\nu) \in V$ and $d(x_{n_\nu} \pi (t_{n_\nu} - \rho_\nu), z) \leq 2\nu^{-1}$. We will show that by choosing U_k small enough, we can arrange that $\omega(U_k \cap S) = A_k$. In fact, if this is not true, then there is a sequence $\delta_\nu \to 0$ such that $\mathrm{Cl}\, U_{\delta_\nu}(A_k) \cap \mathrm{Cl}\, V = \emptyset$, $U_{\delta_\nu}(A_k) \subset U_{k+1}$ and $\omega(U_{\delta_\nu}(A_k) \cap S) \setminus A_k \neq \emptyset$, where $U_{\delta_\nu}(A_k)$ is the δ_ν-neighborhood of A_k. Using what we have proved thus far, it is easily seen that there are sequences $x_\nu \in U_{\delta_\nu}(A_k)$, $s_\nu > 0$ such that $x_\nu \pi s_\nu \in V$ and $d(x_\nu \pi s_\nu, M_{k+1}) \leq 2\nu^{-1}$. There are sequences $\tau_\nu < s_\nu < \tilde{\tau}_\nu \leq \infty$ such that $x_\nu \pi \tau_\nu \in \partial V$, $x_\nu \pi (\tau_\nu, \tilde{\tau}_\nu) \subset \mathrm{Cl}\, V$ and either $\tilde{\tau}_\nu = \infty$ or $x_\nu \pi \tilde{\tau}_\nu \in \partial V$. Set $\hat{x}_\nu = x_\nu \pi \tau_\nu$. We may assume $\hat{x}_\nu \to \hat{x} \in S$. The invariance of A_k and $x_\nu \to A_k$ easily imply $\tau_\nu \to \infty$, so $\hat{x} \in \omega(U_{k+1} \cap S) = A_{k+1}$. On the other hand, $x_\nu \pi s_\nu \to M_{k+1}$ and the invariance of M_{k+1} imply $\tilde{\tau}_\nu \to \infty$ so $\hat{x} \pi [0, \infty) \subset \mathrm{Cl}\, V$. Therefore $\omega(\hat{x}) \subset M_{k+1}$ and $\hat{x} \in A_{k+1}$. Now this obviously implies $\hat{x} \in M_{k+1}$, a contradiction since $\hat{x} \in \partial V$. Hence, indeed, U_k can be chosen such that $\omega(U_k \cap S) = A_k$, i.e. A_k is an attractor.

3. Step:

$M_j = A_j \cap A^*_{j-1}$: Indeed, if $y \in M_j$, then there is a solution $\sigma: \mathbb{R} \to M_j$ through y and therefore $y \in A_j$. Suppose $y \notin A^*_{j-1}$. Then $\omega(y) \subset A_{j-1}$ and therefore $\omega(y) \subset M_k$ for some $k \leq j-1$. Since $\omega(y) \subset M_j$ we get $\omega(y) \subset M_k \cap M_j = \emptyset$, a contradiction. Hence $M_j \subset A_j \cap A^*_{j-1}$. If $y \in A_j \cap A^*_{j-1}$, then there is a solution $\sigma: \mathbb{R} \to S$ through y such that $\omega^*(\sigma) \subset M_1 \cup \ldots \cup M_j$. From $y \in A^*_{j-1}$ we conclude $\omega(y) \cap (M_1 \cup \ldots \cup M_{j-1}) = \emptyset$ and hence $\omega(y) \subset M_k$ for some $k \geq j$. Now the assumptions of the theorem imply $k=j$ and $\sigma[\mathbb{R}] \subset M_j$, and so $y \in M_j$, completing the proof.

Remark:

Theorem 1.8 together with Proposition II.5.4 immediately imply Proposition 1.6.

3.2 Block pairs and index triples

Isolating blocks and, more generally, index pair capture the essential topological properties of a neighborhood of an isolated invariant set K, without, however, taking into account the internal structure of K.

Given a repeller-attractor pair (A*,A) in K, the appropriate concept to use in the analysis of the flow near A*,A and K is that of a <u>block pair</u> and more generally, an <u>index triple</u>.

Definition 2.1

Let K be a compact isolated invariant set in X and (A*,A) be a repeller-attractor pair in K. A pair $<B_1,B_2>$ of subsets of X is called a <u>block pair</u> for (A*,A) relative to K, if the following properties hold:

(i) B_1 is an isolating block for A*, B_2 is an isolating block for A and $B:=B_1 \cup B_2$ is an isolating block for K.

(ii) $B_1 \cap B_2 \subset B_1^- \cap B_2^+$.

Remark:

Property (ii) means that every point $x \in B_1 \cap B_2$ lies in $\partial B_1 \cap \partial B_2$ and x is a strict-egress or a bounce-off point for B_1 while it is a strict ingress or bounce-off point for B_2.

Thus we obtain the following picture:

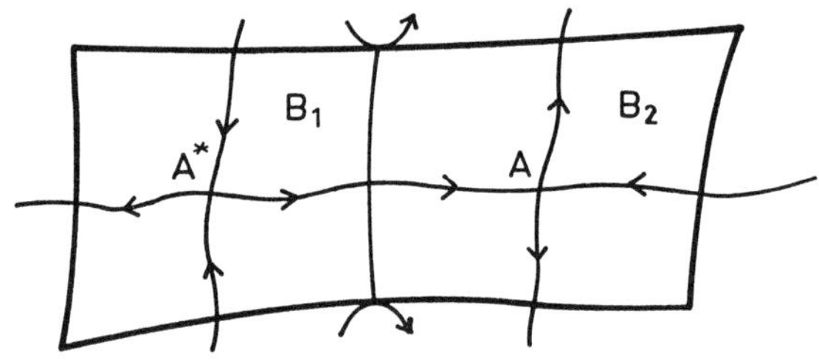

Figure 6

Just as an isolating block gives rise to an index pair, block pair gives rise to an <u>index triple</u>:

Definition 2.2

Let K be a compact isolated invariant set and (A*,A) be a repeller-attractor pair in K. Then the triple $<N_1,N_2,N_3>$ with $N_1 \supset N_2 \supset N_3$ is called an <u>index triple for</u> (A*,A) <u>relative to</u> K if the following properties hold:

(1) N_1 is an isolating neighborhood of K and $<N_1,N_3>$ is an index pair in N_1.

(2) N_2 is an isolating neighborhood of A and $<N_2,N_3>$ is an index pair in N_2.

(3) If $U \subset X$ is any open set with $A \subset U \cap N_1 \subset N_2$, then $N_1 \setminus U$ is an isolating neighborhood of A^* and $\langle N_1 \setminus U,\ N_2 \setminus U \rangle$ is an index pair in $N_1 \setminus U$.
We shall say that an index triple $\langle N_1, N_2, N_3 \rangle$ for (A^*, A), relative to K, satisfies the <u>cofibration property</u> if all the inclusions $N_2 \subset N_1$, $N_3 \subset N_2$ and $N_2 \setminus U \subset N_1 \setminus U$ (where U is as in (3) of Definition 2.2) are cofibrations.
We will now show how block pairs define index triples.

<u>Theorem 2.3</u>

Let K, A, A be as in Definition 2.1 and $\langle B_1, B_2 \rangle$ be a block pair for (A^*, A) relative to K. Let $B := B_1 \cup B_2$. Suppose π does not explode in B. Then the triple $\langle B, B_2 \cup B^-, B^- \rangle$ is an index triple for (A^*, A), relative to K, satisfying the cofibration property.*

<u>Proof</u>:

Let $N_1 = B$, $N_2 = B_2 \cup B^-$, $N_3 = B^-$. Then $N_1 \supset N_2 \supset N_3$.

<u>1. Step</u>

Clearly N_1 is an isolating neighborhood of K, and $\langle N_1, N_3 \rangle$ is an index pair in N_1. Moreover, since π does not explode in $B = N_1$, Theorem I.3.7 implies that the inclusion $N_3 \subset N_1$ is a cofibration.

<u>2. Step</u>

Clearly N_2 is an isolating neighborhood of A, and since $A \subset (\operatorname{Int} B_2) \setminus B^-$, it follows that $A \subset \operatorname{Int}((B_2 \cup B^-) \setminus B^-) = \operatorname{Int}(N_2 \setminus N_3)$. Moreover, N_3 is obviously N_2-positively invariant. Now let $x \in N_2$ with $x\pi t \notin N_2$ for some $t < \omega_x$. Set $\tau = \sup\{s < \omega_x \mid x\pi[0,s] \subset N_2\}$. Then $y := x\pi\tau$ is defined and we claim that $y \in B^-$. In fact, if $y \notin B^-$, then clearly $y \in \partial B_2$. Since B_2 is an isolating block, it follows that $y \notin B_2^i$, i.e. $y \in B_2^-$. Thus for a small $\varepsilon > 0$ we have $y\pi(0,\varepsilon) \cap B_2 = \emptyset$. If $y \notin B_1$, then we may assume that $y\pi[0,\varepsilon) \cap B_1 = \emptyset$, i.e. $y\pi(0,\varepsilon) \cap B = \emptyset$, so $y \in B^-$, a contradiction to our assumption. Hence $y \in B_1$ and so $y \in B_1 \cap B_2 \subset B_1^- \cap B_2^+$. Taking ε smaller if necessary, it follows that $y\pi(0,\varepsilon) \cap B_1 = \emptyset$, i.e. $y\pi(0,\varepsilon) \cap B = \emptyset$, and so $y \in B^-$, again a contradiction.
This shows that $y \in B^-$ and completes the proof that $\langle N_2, N_3 \rangle$ is an index pair in N_2.
To show that the inclusion $N_3 \subset N_2$ is a cofibration we use results from section 1.3.
Let U and $H: U \times [0,1] \to B$ be as in the proof of Theorem I.3.7. Let $\tilde{U} = U \cap (B_2 \cup B^-) = U \cap N_2 = N_2 \setminus A^+(B)$. Obviously $\tilde{U}$ is open in N_2 and $N_3 = B^- \subset \tilde{U}$.
Let us prove that H maps $\tilde{U} \times [0,1]$ into N_2. In fact if this is not true

then for some x there exists an $\alpha, 0 \le \alpha < 1$, such that

$$\alpha = \sup\{t \in [0,1] \,|\, x\pi(s\cdot s_B(x)) \in N_2 \quad \text{for } s \in [0,t]\} .$$

It follows that

$$x\pi(\alpha\cdot s_B(x)) \in N_2 \setminus B^- \subset B_2 .$$

Consequently for some small $\varepsilon>0$ with $\alpha+\varepsilon<1$,

$$x\pi(t\cdot s_B(x)) \in \text{Int } B \quad \text{for } t \in (\alpha,\alpha+\varepsilon) .$$

There is a sequence $t_n \in (\alpha,\alpha+\varepsilon)$ $t_n\to\alpha$ as $n\to\infty$ such that $x\pi(t_n\cdot s_B(x)) \notin N_2$. Thus $x\pi(t_n\cdot s_B(x)) \in B_1\setminus B_2$. It follows that $x\pi(\alpha\cdot s_B(x)) \in B_1 \cap B_2 \subset B_1^- \cap B_2^+$. But $x\pi(\alpha\cdot s_B(x)) \in B_1^-$ implies that $x\pi(t_n\cdot s_B(x)) \notin B_1$ for all n sufficiently large, a contradiction. This contradiction proves that H maps $\tilde{U}\times[0,1]$ into N_2. Let $\tilde{H}:\tilde{U}\times[0,1]\to N_2$ be the corresponding restriction of H. Then an application of Proposition I.3.6 to $\tilde{U}$ and $\tilde{H}$ implies that the inclusion $N_3 \subset N_2$ is a cofibration.

3. Step

Let V be any open set with $A \subset V \cap N_1 \subset N_2$. We will show that $N_1\setminus V$ is an isolating neighborhood of A^*, $\langle N_1\setminus V,\ N_2\setminus V\rangle$ is an index pair in $N_1\setminus V$ and the inclusion $N_2\setminus V \subset N_1\setminus V$ is a cofibration.
Let K^* be the largest invariant set in $N_1\setminus V$. Then $K^* \subset K$. We claim that $K^*=A^*$. In fact, if $x \in K^*\setminus A^*$, then, by Theorem 1.4 $\omega(x) \subset A \subset V$. Since $N_1\setminus V$ is closed, $\omega(x) \subset N_1\setminus V$, a contradiction. This proves that $K^* \subset A^*$. Let $x \in A^*$. Then $x \in \text{Int } B_1 \subset N_1$. Assuming $x \in V$ we get $x \in V\cap N_1 \subset N_2 = B_2 \cup B^-$. Now obviously $x \notin B^-$, so $x \in (\text{Int } B_1)\cap B_2=\emptyset$, a contradiction. Hence, $x \in N_1\setminus V$, i.e. $A^* \subset N_1\setminus V$ which proves that $K^*=A^*$. Moreover, $A^* \subset \text{Int } B_1 \subset \text{Int } B_1\setminus(B_2\cup B^-) \subset (N_1\setminus V)\setminus(N_2\setminus V)$. Hence $A^* \subset \text{Int}((N_1\setminus V)\setminus(N_2\setminus V))$.

We claim that $N_2\setminus V$ is $N_1\setminus V$-positively invariant. In fact, let $x \in N_2\setminus V$ and $x\pi[0,t] \subset N_1\setminus V$. Suppose $x\pi t \notin N_2$. Define τ as in Step 2. Then $\tau<t$ and $x\pi\tau \in B^-$ contradicting the fact that $x\pi[0,t] \subset B$. This proves the claim.
Let us now verify the "exit ramp" property of $N_2\setminus V$. Let $x \in N_1\setminus V$ be such that $x\pi t \notin N_1\setminus V$ for some $t>0$. Let $s=s_{N_1\setminus V}(x)$, and $y=x\pi s$. We have to show that $y \in N_2\setminus V$. We have the following three possible cases (as $y \in (N_1\setminus V)\setminus\text{Int }(N_1\setminus V)$):

1. $y \in (\mathrm{Int}\, N_1) \cap \partial V$:
Then there exists a sequence $x_n \to y$ with $x_n \in V$ and $x_n \in \mathrm{Int}\, N_1$. Thus $x_n \in (V \cap N_1) \subset N_2$, and so $y \in N_2 \setminus V$.

2. $y \in \partial N_1 \setminus \mathrm{Cl}\, V$:
Then $y \in B^-$ for otherwise $y \in B^i$ and so $y\pi[0,\varepsilon] \subset B \setminus \mathrm{Cl}\, V \subset N_1 \setminus V$ for some small $\varepsilon > 0$ contradicting the definition of y.
Thus $y \in B^- \setminus V \subset N_2 \setminus V$.

3. $y \in \partial N_1 \cap \partial V$:
Then either $y \in B^-$ and so $y \in B^- \setminus V \subset N_2 \setminus V$, or else $y \in B^i$. In the latter case $y\pi[0,\delta] \subset B = N_1$ for some $\delta > 0$. The definition of y implies the existence of a sequence $0 < s_1 < \delta$ with $s_n \to 0$ and $y\pi s_n \in (N_1 \cap V) \subset N_2$. This implies $y \in N_2$ so $y \in N_2 \setminus V$.
To show that $N_2 \setminus V \subset N_1 \setminus V$ is a cofibration, define $U = (N_1 \setminus V) \setminus A^+(B_1)$ and let $H: U \times [0,1] \to N_1 \setminus V$ be defined as

$$H(x,t) = \begin{cases} x\pi(t \cdot s_{B_1}(x)) & \text{if } x \in B_1 \\ x & \text{otherwise} \end{cases}$$

We will verify the assumptions of Proposition I.3.6. Let us first prove that H is well-defined, i.e. that $H(x,t) \in N_1 \setminus V$ for $(x,t) \in U \times [0,1]$. This is clear if $x \notin B_1$ or if $x \in B_1^-$. Let $x \in B_1 \setminus B_1^-$ and $0 \leq t < 1$. Suppose $H(x,t) \in V$. Then $y = x\pi(t \cdot s_{B_1}(x) \in B_1 \cap V \subset N_1 \cap V \subset N_2 = B_2 \cup B^-$. Since $t \cdot s_{B_1}(x) < s_{B_1}(x) \leq s_B(x)$, we obtain $y \notin B^-$. Thus $y \in B_1 \cap B_2 \subset B_1^-$, a contradiction. This proves that $H(x,t) \in N_1 \setminus V$ if $t < 1$, $x \in B_1$. By continuity of π, $H(x,1) \in N_1 \setminus V$. Hence, indeed, H is well-defined. Furthermore, U is open in $N_1 \setminus V$. Let $x \in N_2 \setminus V = (B_2 \cup B^-) \setminus V$ and suppose that $x \in A^+(B_1) \subset B_1 \subset B$. Then either $x \in B_2$, and so $x \in B_1 \cap B_2 \subset B_1^-$ a contradiction, or else $x \in B^-$, again a contradiction.
This proves that $N_2 \setminus V \subset U$.
Let us prove that H is continuous. To this end, we only have to consider the case $(x_n,t_n) \to (x,t)$ with $x_n \in U \setminus B_1$, $x \in B_1$. In this case $H(x_n,t_n) = x_n$. Now $U \setminus B_1 \subset B_2$ and so $x_n \in B_2$, i.e. $x \in B_1 \cap B_2 \subset B_1^-$. Thus $s_{B_1}(x) = 0$ and so $H(x,t) = x\pi(t \cdot s_{B_1}(x)) = x$.
This proves the continuity of H.
Next we claim that $H(x,t) = x$ for all $x \in N_2 \setminus V$ and $t \in [0,1]$. Indeed, we may assume $x \in B_1$. If $x \in B_2$ then $x \in B_1 \cap B_2 \subset B_1^-$ and so $H(x,t) = x$. Else if $x \notin B_2$ then $x \in B^-$ and so $0 = s_B(x) \geq s_{B_1}(x) \geq 0$. Thus $s_{B_1}(x) = 0$ and, again, $H(x,t) = x$.
Finally, let us show that $H(x,1) \in N_2 \setminus V$ for all $x \in U$. If $x \in B_1$, then

$H(x,t)=x\pi s_{B_1}(x)=y \in B_1^- \setminus V$. If $y \notin B^-$, then for some $\varepsilon>0$ $y\pi(0,\varepsilon) \in B\setminus B_1 \subset B_2$. Thus $y \in B_2 \setminus V \subset N_2 \setminus V$. If $y \in B^-$ then again, $y \in B^- \setminus V \subset N_2 \setminus V$. If $x \notin B_1$, then $x \in B_2 \setminus V \subset N_2 \setminus V$ and $H(x,1)=x$. Again $H(x,1) \in N_2 \setminus V$.
Now Proposition I.3.6 implies that the inclusion $N_2 \setminus V \subset N_1 \setminus V$ is a cofibration, completing the proof of the theorem.
We will now prove a basic existence theorem for block pairs:

Theorem 2.4

Let $(\pi,K) \in \mathcal{S}$ *and* $\hat{N}$ *be a strongly* π*-admissible isolating neighborhood of* K. *Let* (A^*,A) *be a repeller-attractor pair in* K.
Then there exists a block pair (B_1,B_2) *for* (A^*,A) *relative to* K, *with* $B:=B_1 \cup B_2 \subset \hat{N}$.

Proof:

K is compact, of course. Suppose first that $A=\emptyset$ or $A^*=\emptyset$. If $A=\emptyset$, then $A^*=K$ by the definition of a dual repeller. If $A^*=\emptyset$ and $x \in K\setminus A$ then for any full solution $\sigma:\mathbb{R}\to K$ through x, $\omega^*(\sigma)\subset A^*$ by Theorem 1.4, and $\omega^*(\sigma)$ is nonempty, a contradiction. Thus $K=A$. Hence $(A^*,A)=(K,\emptyset)$ or $(A^*,A)=(\emptyset,K)$. Let $B\subset N$ be an isolating block for K. In the first case take $B_1=B$, $B_2=\emptyset$ whereas in the second case $B_1=\emptyset$, $B_2=B$. (B_1,B_2) and B obviously satisfy the requirements of Definition 2.1.
Now assume $A^*\neq\emptyset$ and $A\neq\emptyset$. Since $A^*\cap A=\emptyset$, there exists an open set $U\subset X$ with $A^*\subset U\subset\hat{N}$ and $N\cap A=\emptyset$, where $N:=\mathrm{Cl}\ U$. Theorem 1.4(ii) easily implies that N is an isolating neighborhood of A^* and $N\subset\hat{N}$ implies that $\hat{N}$ is strongly π-admissible.
Define $g^+:U\to\mathbb{R}^+$ as in (v) following the statement of Theorem I.5.1, with K replaced by A^*. By Proposition I.5.2, (2), there are open sets V and W satisfying $A^*\subset V\subset \mathrm{Cl}\ V\subset W\subset \mathrm{Cl}\ W\subset U$ and such that $g^+|\mathrm{Cl}\ W$ is continuous.
Define $G_\varepsilon=\mathrm{Cl}\ V_\varepsilon$ where

$$V_\varepsilon = \{y\in V \mid g^+(y)<\varepsilon\} .$$

Since $A^*\subset V_\varepsilon$ and V_ε is open in X, $A^*\subset \mathrm{Int}\ G_\varepsilon$. Now we have

Lemma 2.5

There is an isolating neighborhood $\tilde{N}\subset\hat{N}$ *of* K *such that*

$$\tilde{N}\cap\partial G_\varepsilon\subset V$$

for some $\varepsilon>0$.

Proof of the lemma:

First we claim that $(K\setminus A)\cap\partial G_\varepsilon\subset V$ for some $\varepsilon>0$. In fact, otherwise the compactness of K implies that there are sequences $x_n\to x\in K$, $\varepsilon_n\to 0$ with $x_n\in(K\setminus A)\cap\partial G_{\varepsilon_n}\cap\partial V$. Since g^+ is continuous on Cl W, it follows that $x\in K\cap\partial V$ and $g^+(x)=0$. Moreover, $x\notin A$ since $\partial V\cap A=\emptyset$. Proposition I.5.2 (2) implies that $x\in A^+(N)$ and so $\omega(x)\subset A^*$ which by Theorem 1.4 means that $x\in A^*$, a contradiction since $A^*\cap\partial V=\emptyset$.
Let $\varepsilon>0$ be as in the claim. If the lemma is not true, then there is a sequence $y_n\to y\in K$ such that $y_n\in\partial G_\varepsilon\cap\partial V$. Thus $y\in\partial G_\varepsilon\cap\partial V\cap K$ and so $y\in(K\setminus A)\cap\partial G_\varepsilon\subset V$ a contradiction. The lemma is proved.
Lemma 2.5 and Theorem I.5.1 imply that there is an isolating block B for K with $B\subset\tilde{N}$, i.e. $B\cap\partial G_\varepsilon\subset V$ for some $\varepsilon>0$.
Choose such ε and B and define $B_1=B\cap G_\varepsilon$ and $B_2=\mathrm{Cl}\,(B\setminus G_\varepsilon)$.
We will show that (B_1,B_2) is a block pair for (A^*,A) relative to K.

Lemma 2.6

B_1 is an isolating block for A^*, with $B_1^-=\partial B_1\cap(B^-\cup\partial G_\varepsilon)$.

Proof of the lemma:

Let K_1 be the largest invariant set in B_1. Since $A^*\subset\mathrm{Int}\,G_\varepsilon\cap\mathrm{Int}\,B$, we have $A^*\subset K_1$. $K_1\subset G_\varepsilon\subset\mathrm{Cl}\,V$ implies that $K_1\cap A=\emptyset$ and so by Theorem 1.4 $K_1=A^*$. Thus B_1 is an isolating neighborhood of A^*.
In order to verify that B_1 is an isolating block, let $x\in\partial B_1$ and let $\sigma:[-\delta_1,\delta_2]\to X$ be a solution through x, where $-\delta_1\leq 0<\delta_2$.
(a) Assume $x\in\mathrm{Int}\,G_\varepsilon$. If δ_1 and δ_2 are small, and $-\delta_1\leq t\leq\delta_2$, then $\sigma(t)\in\mathrm{Int}\,B$ implies $\sigma(t)\in\mathrm{Int}\,B_1$ and $\sigma(t)\in X\setminus B$ implies $\sigma(t)\in X\setminus B_1$.
Therefore $x\in B^i$ (resp. $x\in B^e$, resp. $x\in B^b$) implies $x\in B_1^i$ (resp. $x\in B_1^e$, resp. $x\in B_1^b$).
(b) Assume $x\in\partial G_\varepsilon$. Then $g^+(x)=\varepsilon$. In fact, $g^+(x)\leq\varepsilon$ by continuity of g^+ on Cl W. Since $B\cap\partial G_\varepsilon\subset V$ we have $x\in V$ and so $g^+(x)<\varepsilon$ would imply $x\in V_\varepsilon\subset\mathrm{Int}\,G_\varepsilon$ a contradiction.
If δ_1 and δ_2 are sufficiently small, $\sigma(t)\in V$ and by Proposition I.5.2, (2) we obtain that $g^+(\sigma(t))<\varepsilon$ if $t<0$, $=\varepsilon$ for $t=0$, $>\varepsilon$ if $t>0$. Consequently $\sigma[[-\delta_1,0)]\subset\mathrm{Int}\,G_\varepsilon$ and $\sigma[(0,\delta_2]]\subset X\setminus G_\varepsilon$. Hence, $x\in\mathrm{Int}\,B\cup B^e$ (resp. $x\in B^i\cup B^b$) implies $x\in B_1^e$ (resp. $x\in B_1^b$). Altogether we obtain

$$\partial B_1\cap(\mathrm{Int}\,B\cup B^e)\subset B_1^e \qquad (2)$$
$$\partial B_1\cap B^b\subset B_1^b \qquad (3)$$
$$\partial B_1\cap B^i\cap\mathrm{Int}\,G_\varepsilon\subset B_1^i \qquad (4)$$
$$\partial B_1\cap B^i\cap\partial G_\varepsilon\subset B_1^b \qquad (5).$$

Relations (2)-(5) imply that B_1 is an isolating block (for A^*) and $B_1^-=\partial B_1\cap(\partial G_\varepsilon\cup B^-)$.

Lemma 2.7

B_2 is an isolating block for A and $B_2^-=B^-\cap\partial B_2$.

Proof of the lemma:

Let K_2 be the largest invariant set in B_2. Since $A\subset\text{Int }B\cap(X\setminus G_\varepsilon)\subset\text{Int }B_2$ we find $A\subset K_2$. If $A\neq K_2$, then there is an $x\in K_2\setminus A$ and a full solution $\sigma:\mathbb{R}\to K_2\subset K$ through x. By Theorem 1.4, $\omega^*(\sigma)\subset A^*\cap K_2\subset A^*\cap B_2=\emptyset$, a contradiction. Hence $K_2=A$ and so B_2 is an isolating neighborhood of A. Let $x\in\partial B_2$ and $\sigma:[-\delta_1,\delta_2]\to X$ be a solution through x, with $-\delta_1\leq 0<\delta_2$.

(a) Assume $x\in X\setminus G_\varepsilon$. Then $x\in\partial B$. If δ_1,δ_2 are small, then $\sigma(t)\in\text{Int }B$ implies $\sigma(t)\in\text{Int }B_2$ and $\sigma(t)\in X\setminus B$ implies $\sigma(t)\in X\setminus B_2$. Thus $x\in B^i$ (resp. $x\in B^e$, resp. $x\in B^b$) implies that $x\in B_2^i$ (resp. $x\in B_2^e$, resp. $x\in B_2^b$).

(b) Let $x\in G_\varepsilon$. Then $x\in\partial G_\varepsilon$ and as in the proof of Lemma 2.6 we obtain (for δ_1,δ_2 small) that $\sigma[[-\delta_1,0)]\subset\text{Int }G_\varepsilon$ and $\sigma[(0,\delta_2])\subset X\setminus G_\varepsilon$. Consequently $x\in\text{Int }B\cup B^i$ (resp. $x\in B^e\cup B^b$) implies $x\in B_2^i$ (resp. B_2^b). Altogether we obtain

$$\partial B_2\cap(\text{Int }B\cup B^i)\subset B_2^i \tag{6}$$
$$\partial B_2\cap B^b\subset B_2^b \tag{7}$$
$$\partial B_2\cap B^e\cap(X\setminus G_\varepsilon)\subset B_2^e \tag{8}$$
$$\partial B_2\cap B^e\cap\partial G_\varepsilon\subset B_2^b \tag{9}$$

It follows that B_2 is an isolating block (for A) and $B_2^-=B^-\cap\partial B_2$.
To complete the proof of the theorem, it is now only necessary to show that $B_1\cap B_2\subset B_1^-\cap B_2^+$. Clearly $B_1\cap B_2\subset\partial G_\varepsilon\cap B$. Let us show that $B_1\cap B_2\subset\partial B_1\cap\partial B_2$. In fact, since $\text{Int }B_1\subset\text{Int }G_\varepsilon$, we get $(B_1\cap B_2)\cap\text{Int }B_1=\emptyset$, i.e. $B_1\cap B_2\subset\partial B_1$. If $x\in(B_1\cap B_2)\cap\text{Int }B_2$ then $x\in\partial G_\varepsilon$ implies that there is a sequence $x_n\in V_\varepsilon\subset\text{Int }G_\varepsilon$ with $x_n\to x$, so $x_n\in\text{Int }B_2\cap\text{Int }G_\varepsilon$ for all $n\geq n_0$, and thus $x_n\notin\text{Cl }(B\setminus G_\varepsilon)=B_2$, a contradiction.
Therefore, indeed, $B_1\cap B_2\subset\partial B_1\cap\partial B_2$.
Let $x\in B_1\cap B_2$. If $x\notin B_1^-$, then $x\in B_1^i$ and so by (2)-(5) $x\in\text{Int }G_\varepsilon$ which contradicts the fact that $x\in\partial G_\varepsilon$. Hence $x\in B_1^-$.
If $x\notin B_2^+$, then $x\in B_2^e\setminus B_2^b$, and so by (6)-(9), $x\in X\setminus G_\varepsilon$, again a contradiction.
The theorem is proved completely.

3.3 A Morse equation

Let $(\pi,K) \in S(X)$ and $(M_1,\ldots,M_n)$ be a Morse decomposition of K. In this section we will prove a fundamental equation relating the Betti-numbers of $h(M_i)$ with the Betti-numbers of $h(K)$, thus extending classical results from Morse theory.

For the rest of this section, H_q, resp. H^q, will denote an arbitrary homology, resp. cohomology theory with coefficients in an R-module M where R is an integral domain (Artin and Braun [1]). We tacitly assume that H_q, resp. H^q, is defined on all spaces and pairs of spaces used below.

This means that H_q (resp. H^q) should be defined for pairs (A,B) and (A/B, [B]) where $B \subset A$, B,A are closed in X and A is strongly π-admissible. In particular if π is the local flow on $U \subset \mathbb{R}^n$ generated by an ordinary differential equation, H_q, resp. H^q need only be defined on pairs of compact spaces. We will use this remark in the proofs to come.

We also assume that H_q (resp. H^q) is $\{0\}$ for $q<0$.

Definition 3.1

Let E be an R-module and $\tilde{E}=E\otimes_R K$, where K is the quotient field of R (see e.g. Artin and Braun [1]). Then we define

$$\text{rank } E = \begin{cases} \dim \tilde{E} & , \text{ if } \dim \tilde{E} \text{ is finite} \\ \infty & , \text{ otherwise} \end{cases}$$

Now we have the following

Lemma 3.2

Let E,F,G be R-modules and the sequence

$$E \xrightarrow{f} F \xrightarrow{g} G$$

of R-homomorphisms be exact, i.e. $\operatorname{Im} f=\ker g$.
Then

$$\text{rank } F = \text{rank } (\operatorname{Im} f) + \text{rank } (\operatorname{Im} g) \quad .$$

Remark:

Of course, we adopt the usual convention that $\infty+a=\infty$ for $a \in \mathbb{R}^+ \cup \{\infty\}$.

Proof:

Let K be the quotient ring of R. Since the functor $\otimes_R K$ preserves exactness (see Artin and Braun [1]) there is a sequence

$$E\otimes_R K \xrightarrow{f\otimes 1} F\otimes_R K \xrightarrow{g\otimes 1} G\otimes_R K$$

of K-vector spaces and K-homomorphisms with $\mathrm{Im}(f\otimes 1)=\ker(g\otimes 1)$. Rank E is, by definition, equal to $\dim(E\otimes_R K)$, similarly for G. Moreover, clearly $\mathrm{Im}(f\otimes 1)=(\mathrm{Im}\, f)\otimes_R K$ with a similar expression for g. Therefore it is enough to prove the lemma if R is a field, i.e. all the spaces are vector spaces. But then, trivially

$$F \cong \ker g\oplus(F/\ker g) \cong \ker g\oplus \mathrm{Im}\, g \cong \mathrm{Im}\, f\oplus \mathrm{Im}\, g .$$

From this the lemma follows immediately.

Definition 3.3

If $(\pi,S) \in \mathcal{S}(X)$, then the (formal) Poincaré-polynomial $p(t,h(S))$ with respect to H_q (or H^q) is defined as

$$p(t,h(S)) = \sum_{q=0}^{\infty} \beta_q(h(S))\cdot t^q$$

$$(\text{resp. } p(t,h(S)) = \sum_{q=0}^{\infty} \beta^q(h(S))\cdot t^q) .$$

Here

$$\beta_q(h(S)) = \mathrm{rank}\, H_q(h(S)),$$

$$\beta^q(h(S)) = \mathrm{rank}\, H^q(h(S)) .$$

Note that some coefficients of $p(t,h(S))$ may be equal to ∞.

The following lemma holds:

Lemma 3.4

For every $j=1,\ldots,n$ there exists and index triple $\langle N_1^j,N_2^j,N_3^j\rangle$ for the repeller-attractor pair (M_j,A_{j-1}) relative to A_j such that for $q \in \mathbb{Z}$

$$H_q(h(M_j)) \cong H_q(N_1^j,N_2^j) \cong H_q(N_1^j\setminus U,N_2^j\setminus U) .$$

$$H_q(h(A_j)) \cong H_q(N_1^j,N_3^j)$$

$$H_q(h(A_{j-1})) \cong H_q(N_2^j,N_3^j) .$$

<u>Here</u>, U <u>and</u> V <u>are open sets in</u> X <u>with</u> Cl $U\subset V$ <u>and</u> $N_1^j\cap V\subset N_2^j$. <u>An analogous statement holds for cohomology</u>.

<u>Proof</u>:

By Theorem 1.7 (M_j,A_{j-1}) is, indeed, a repeller-attractor pair in A_j. By Theorem 2.4 there exists a block pair $\langle B_1,B_2\rangle$ for (M_j,A_{j-1}) relative to A_j. Theorem 2.3 implies that $\langle B,B_2\cup B^-,B^-\rangle$, $B:=B_1\cup B_2$ is an index triple for (M_j,A_{j-1}) relative to A_j, satisfying the cofibration property. Set $N_1^j=B$, $N_2^j=B_2\cup B^-$, $N_3^j=B^-$.
By the excision property of homology we have for all $q\in Z$

$$H_q(N_1^j\setminus U,\ N_2^j\setminus U)\ \cong\ H_q(N_1^j,N_2^j) \tag{1}$$

Proposition I.10.9 together with (1) imply that

$$H_q(h(M_j))\cong H_q((N_1^j\setminus U)/(N_2^j\setminus U),\{[N_2^j\setminus U]\})\cong H_q(N_1^j\setminus U,\ N_2^j\setminus U)\cong H_q(N_1^j,N_2^j)$$

$$H_q(h(A_j))\cong H_q(N_1^j/N_3^j,\{[N_3^j]\})\cong H_q(N_1^j,N_3^j)$$

$$H_q(h(A_{j-1}))\cong H_q(N_2^j/N_3^j,\{[N_3^j]\})\cong H_q(N_2^j,N_3^j)\ .$$

The lemma is proved for homology. The proof for cohomology is analogous.
By well-known results from homology theory, there exists an exact sequence of homology modules:

$$\to H_{q+1}(N_1^j,N_2^j)\xrightarrow{\gamma_q^j} H_q(N_2^j,N_3^j)\xrightarrow{\varepsilon_q^j} H_q(N_1^j,N_3^j)\xrightarrow{\alpha_q^j} H_q(N_1^j,N_2^j)\xrightarrow{\gamma_{q-1}^j} \tag{2}$$

with an analogous (reversed) sequence for cohomology.
Here, (N_1^j,N_2^j,N_3^j) are as in Lemma 3.4.
Therefore, applying Lemma 3.2 to the sequence (2), we obtain:

$$\operatorname{rank} H_q(h(A_{j-1}))\ =\ \operatorname{rank}(\operatorname{Im}\gamma_q^j)+\operatorname{rank}(\operatorname{Im}\varepsilon_q^j) \tag{3a}$$

$$\operatorname{rank} H_q(h(A_j))\ =\ \operatorname{rank}(\operatorname{Im}\varepsilon_q^j)+\operatorname{rank}(\operatorname{Im}\alpha_q^j) \tag{3b}$$

$$\operatorname{rank} H_q(h(M_j))\ =\ \operatorname{rank}(\operatorname{Im}\alpha_q^j)+\operatorname{rank}(\operatorname{Im}\gamma_{q-1}^j) \tag{3c}$$

thus

$$\operatorname{rank} H_q(h(A_{j-1}))+\operatorname{rank} H_q(h(M_j))=$$

$$=\operatorname{rank} H_q(h(A_j))+\operatorname{rank}(\operatorname{Im}\gamma_{q-1}^j)+\operatorname{rank}(\operatorname{Im}\gamma_q^j) \tag{4}$$

Now we claim that

$$\sum_{j=1}^{n} \operatorname{rank} H_q(h(M_j)) = \operatorname{rank} H_q(h(K)) + \sum_{j=1}^{n} (d_{q-1}^j + d_q^j) \tag{5}$$

where $d_q^j := \dim \gamma_q^j$.

In fact, since $A_0 = \emptyset$ and so $H_q(h(A_0)) \equiv \{0\}$, and since $A_n = K$ we obtain from (4)

$$\sum_{j=1}^{n-1} \operatorname{rank} H_q(h(A_j)) + \sum_{j=1}^{n} \operatorname{rank} H_q(h(M_j)) =$$

$$= \sum_{j=1}^{n-1} \operatorname{rank} H_q(h(A_j)) + \operatorname{rank} H_q(h(K)) + \sum_{j=1}^{n} (d_{q-1}^j + d_q^j).$$

Thus, if $\sum_{j=1}^{n-1} \operatorname{rank} H_q(h(A_j))$ is finite, formula (5) follows immediately. Therefore assume that $\sum_{j=1}^{n-1} \operatorname{rank} H_q(h(A_j)) = \infty$. Let ν be the smallest index such that $\operatorname{rank} H_q(h(A_\nu)) = \infty$. Then $\nu \geq 1$, and by (4), $\operatorname{rank} H_q(h(M_\nu)) = \infty$. Thus the left-hand side of (5) is equal to ∞. Assume $\sum_{j=1}^{n} (d_{q-1}^j + d_q^j) < \infty$. Then, using (4) and proceeding recursively, we obtain $\operatorname{rank} H_q(h(A_j)) = \infty$ for all $j = \nu, \dots, n$. Thus $\operatorname{rank} H_q(h(K)) = \infty$, thus the right-hand side of (5) is equal to ∞, and so (5) follows. If $\sum_{j=1}^{n} (d_{q-1}^j + d_q^j) = \infty$, then (5) follows again, and the claim is proved.

Now we obtain the following important

<u>Theorem 3.5</u> (<u>Morse equation</u>)

<u>Let</u> $(\pi, K) \in \mathcal{S}(X)$ <u>and let</u> $(M_1, \dots, M_n)$ <u>be a Morse decomposition of</u> K. <u>Then there are formal polynomials</u> $Q_j(t) = \sum_{q=0}^{\infty} d_q^j t^q$, $j = 1, \dots, n$, <u>whose coefficients are nonnegative integers or</u> ∞ <u>and such that</u>

$$\sum_{j=1}^{n} p(t, h(M_j)) = p(t, h(K)) + (1+t) Q(t) \tag{6}$$

<u>where</u> $Q(t) = \sum_{j=1}^{n} Q_j(t)$.

<u>If</u> $Q_j(t) \neq 0$, <u>then there exists a solution</u> $\sigma: \mathbb{R} \to K$ <u>with</u> $\omega^*(\sigma) \subset M_j$ <u>and</u> $\omega(\sigma) \subset M_i$ <u>for some</u> $i < j$.

<u>Proof</u>:

The proof of (6) follows by multiplying both sides of formula (5) by t^q and summing over all $q \in \mathbb{Z}$.

Suppose that $Q_j(t)\neq 0$ and there is no solution $\sigma:\mathbb{R}\to K$ with $\omega^*(\sigma)\subset M_j$ and $\omega(\sigma)\subset M_i$ for some $i<j$. We shall reach a contradiction. Since (M_j,A_{j-1}) are a repeller-attractor pair in A_j, we obtain that $A_j=A_{j-1}\dot{\cup}M_j$. In fact otherwise, Theorem 1.4 would imply the existence of a solution $\sigma:\mathbb{R}\to K$ with $\omega^*(\sigma)\subset M_j$, $\omega(\sigma)\subset A_{j-1}$. Since $A_{j-1}\cap M_j=\emptyset$, we would get that $\omega(\sigma)\subset M_i$ for some $i<j$ (cf. Theorem 1.7).
Therefore, indeed, $A_j=A_{j-1}\dot{\cup}M_j$.
Now (4) implies that $Q_j(t)$ does not depend on the choice of an index triple (N_1^j,N_2^j,N_3^j). Therefore we can choose a triple $(B,B_2\cup B^-,B^-)$ where $B=B_1\cup B_2$, $B_1\cap B_2=\emptyset$, B_1 is an isolating block for M_j, B_2 an isolating block for A_{j-1}. An application of the sequence (2) to this special index triple yields:

$$\xrightarrow{\gamma_q} H_q(B_2\cup B^-,B^-) \xrightarrow{\xi_q} H_q(B_2\cup B_1,B^-) \ .$$

Consider the following sequence of inclusions

$$(B_2,B_2^-) \xrightarrow{e_1} (B_2\dot{\cup}B_1^-,B_2^-\cup B_1^-) \xrightarrow{e_2} (B_2\dot{\cup}B_1,B_2^-\cup B_1^-) \xrightarrow{e_3} (B_2\dot{\cup}B_1,B_2^-\dot{\cup}B_1)$$

and let e_i^* be induced morphisms of the q'th homology groups.
By excision, $(e_3\circ e_2\circ e_1)^*=e_3^*\circ e_2^*\circ e_1^*$ is an isomorphism. By the same reason e_1^* is an isomorphism. Hence $e_3^*\circ e_2^*$ is an isomorphism, so e_2^* is injective. But $e_2^*=\xi_q$ and so $\ker\xi_q=0$. However, $\ker\xi_q=\operatorname{Im}\gamma_q$.
Therefore $\operatorname{Im}\gamma_q\equiv\{0\}$ for all q, i.e. $Q_j(t)\equiv 0$, a contradiction, which completes the proof of the theorem.

Remarks:

An analogous result is valid for cohomology. The proof is the same.

Formula (4) implies that if $q\in Z$ and rank $H_q(h(M_j))<\infty$ for all $j=1,\ldots,n$, then also rank $H_q(h(A_j))<\infty$ and $d_{q-1}^j+d_q^j<\infty$ for all $j=1,\ldots,n$. Moreover, if rank $H_q(h(M_j))=0$ for all $j=1,\ldots,n$, then rank $H_q(h(A_j))=0$ and $d_{q-1}^j+d_q^j=0$ for all $j=1,\ldots,n$.
Note that the Morse equation (6) is valid even if some of the coefficients of the polynomials occurring in (6) are equal to ∞. Thus we do not have to check whether rank $H_q(h(M_j))$ is finite or not before applying the Morse equation.

From these remarks and Theorem 3.5 we obtain

Corollary 3.6

Let $(\pi,K)\in S(X)$ and $(M_1,\dots,M_n)$ be a Morse decomposition of K. Suppose that $h(M_j)=\Sigma^{d_j}$ for some $d_j\geq 0$, $j=1,\dots,n$. Let m_k be the number of the M_i's with $h(M_i)=\Sigma^k$. Then

$$\sum_{q=0}^{\infty} m_q t^q = \sum_{q=0}^{\infty} \beta_q(h(K))t^q+(1+t)Q(t) \tag{7}$$

Q has finite nonnegative coefficients which are zero for all q sufficiently large. A similar result holds for cohomology.

Proof:

This follows noticing that rank $H_q(S^m,s_0)=\delta_{mq}$ where (S^m,s_0) is an m-dimensional pointed sphere and δ_{mq} is the Kronecker delta (cf. Switzer [1]).
Let a_q, $q\geq 0$, be the coefficients of $Q(t)$.
Then $(1+t)Q(t)=\sum_{q=0}^{\infty} a_q(t^q+t^{q+1})=\sum_{q=0}^{\infty}(a_q+a_{q-1})t^q$ where $a_{-1}=0$.
This implies

$$m_q = \beta_q+(a_q+a_{q-1}) \tag{8}$$

for $q\geq 0$, where $\beta_q=\beta_q(h(K))$.
In particular, for every $k\geq 0$

$$\sum_{q=0}^{k}(-1)^q m_q = \sum_{q=0}^{k}(-1)^q\beta_q+(-1)^k a_k. \tag{9}$$

Since $a_q=0$ for q sufficiently large we obtain

$$\sum_{q=0}^{\infty}(-1)^q m_q = \sum_{q=0}^{\infty}(-1)^q\beta_q = \gamma(h(K)) \tag{10}$$

where $\gamma(h(K))$ is the Euler characteristic of $h(K)$. Furthermore, $a_k\geq 0$ implies

$$(-1)^k\sum_{q=0}^{k}(-1)^q m_q \geq (-1)^k\sum_{q=0}^{k}(-1)^q\beta_q \tag{11}$$

for every $k\geq 0$.
Corollary 3.6 and formulas (10), (11) are, in particular, applicable to a gradient flow on a compact differentiable manifold M (without boundary), having only nondegenerate equilibria a_i. In fact, there are only a finite number of such equilibria, $a_1,\dots,a_n$ and they can be ordered to form a Morse decomposition of K=M (see Proposition 1.6).

Moreover, since K=M, B=M is an isolating block with $B^{-}=\emptyset$. Thus h(K) is the homotopy type of $(M\dot{\cup}\{p\},\{p\})$ where $p\notin M$. By excision, $H_q(M\dot{\cup}\{p\},\{p\})\tilde{=}H_q(M)$ and so $\beta_q(h(K))=\beta_q(M)$, $\beta_q(M)$ being the Betti-numbers of the manifold M. Moreover, $h(\{a_i\})=\Sigma^{d_i}$ where d_i is the dimension of the unstable manifold of a_i.

Consequently, in this special case (10), (11) reduce to well-known formulas of classical Morse theory.

In the next section we will look more closely into the relation between the homotopy index and the Morse index for gradient flows on finite- or infinite-dimensional Hilbert-manifolds.

As an illustration of the Morse equation, let us prove the following

Proposition 3.7:

Assume all hypotheses of Corollary II.6.2. *In addition, suppose that all equilibria* u_0 *of the equation*

$$\dot{u}+Au = f(u) \tag{12}$$

are hyperbolic, i.e. re $\sigma(A-f'(u_0))\neq 0$. *Then there are at least two nontrivial equilibria of* (12).

Proof:

By Corollary II.6.2, there are at least two equilibria of (12): $u_1=0$ and $u_2\neq 0$. Suppose these are the only equilibria of (12). Then $(\{u_1\},\{u_2\})$ or $(\{u_2\},\{u_1\})$ form a Morse decomposition of K_∞. By our assumption and Theorem II.3.5, there are nonnegative integers p,r,q such that $p\neq q$ and

$$h(\pi,\{u_1\}) = \Sigma^p$$
$$h(\pi,\{u_2\}) = \Sigma^r$$
$$h(\pi,K_\infty) = \Sigma^q.$$

By the Morse equation (6), there are nonnegative integers $a_{-1},a_0,a_1,\ldots,$ with $a_{-1}=0$ and such that

$$t^p+t^r=t^q+\sum_{j=0}^{\infty}(a_j+a_{j-1})t^j = \sum_{\substack{j=0\\ j\neq q}}^{\infty}(a_j+a_{j-1})t^j+(a_q+a_{q-1}+1)t^q.$$

Since $p\neq q$, we obtain

$$a_q+a_{q-1}+1 = 1 \text{ and so } q=r \text{ and } a_q+a_{q-1} = 0.$$

Hence

$$t^p = \sum_{j=0}^{\infty} (a_j + a_{j-1}) t^j$$

which implies $a_p + a_{p-1} = 1$ and $a_j + a_{j-1} = 0$ for $j \neq p$. However, this is an obvious contradiction which proves the proposition.

We will conclude this section by examining the relation between the homotopy index and the Brouwer mapping degree $d(f,G,0)$.

Theorem 3.8

Assume the following hypotheses:

(1) U is open in $\mathbb{R}^m$ and $G: U \to \mathbb{R}$ is a C^1-function with $g := \nabla G$ locally Lipschitzian.

(2) Ω is open and bounded in $\mathbb{R}^m$ and $N = \mathrm{Cl}\ \Omega$ is an isolating neighborhood of an isolated invariant set K for the equation

$$\dot{u} = g(u). \tag{12}$$

Under these assumptions, $\beta_q = \beta_q(h(K))$ is finite for all $q \geq 0$ and zero for all q sufficiently large. Moreover, $d(g,\Omega,0)$ is defined and

$$d(g,\Omega,0) = (-1)^m \sum_{q=0}^{\infty} (-1)^q \beta_q \ . \tag{13}$$

The same result holds for cohomology.

Proof:

Let π be the local flow generated (on U) by (12). Then π is gradient-like with respect to $-G$. Since $K \cap \partial N = \emptyset$, in particular, $g(x) \neq 0$ for $x \in \partial\Omega$. Thus $d(g,\Omega,0)$ is defined. Choose any sequence $G_\nu: U \to \mathbb{R}$ of C^∞-functions converging to G uniformly on any compact set in U. By results in Section 1.12, for all ν sufficiently large, N is an isolating neighborhood of an invariant set K_ν relative to the equation

$$\dot{u} = g_\nu(u) := \nabla G_\nu(u) \ . \tag{14}$$

Moreover, by the homotopy invariance of both the homotopy index and the degree

$$h(K_\nu) = h(K) \tag{15}$$

and

$$d(g_\nu,\Omega,0) = d(g,\Omega,0) \tag{16}$$

for such ν.

Therefore we assume w.l.o.g. that G is a C^∞-function. By Corollary 6.8 in Milnor [1] there is a sequence of C^∞-functions $\tilde{G}_\nu: U \to \mathbb{R}$ having only nondegenerate critical points, converging to G uniformly on compact sets in U. Using the same argument we can therefore assume that G has only nondegenerate critical points. Let $A_1, \ldots, a_n$ be the critical points of G in N. We may assume that $(\{a_1\}, \ldots, \{a_n\})$ are ordered in such a way as to form a Morse decomposition of K.
Now $h(\{a_K\}) = \sum^{d_K}$ where d_K is the number of positive eigenvalues of the Hessian $(\nabla^2_{ij} G(a_K))_{ij}$, counted with their multiplicity.
By (10)

$$\sum_{k=1}^{n} (-1)^{d_k} = \sum_{q=0}^{\infty} m_q (-1)^q = \sum_{q=0}^{\infty} (-1)^q \beta_q \;. \tag{17}$$

Now by the definition of the degree

$$d(g,\Omega,0) = \sum_{k=1}^{n} \operatorname{sign} \det(\nabla^2_{ij} G(a_k))_{ij} = \sum_{k=1}^{n} (-1)^{m-d_k} = \sum_{k=1}^{n} (-1)^{m-d_k} (-1)^{2d_k} =$$

$$= (-1)^m \sum_{k=1}^{n} (-1)^{d_k} = (-1)^m \cdot \sum_{q=0}^{\infty} (-1)^q \beta_q \quad .$$

The theorem is proved.

Corollary 3.9

Under the assumptions of Theorem 3.8, if $h(K)=0$ then $d(g,\Omega,0)=0$.
Corollary 3.9 proves the remark preceding Proposition II.8.6.

3.4 The homotopy index and Morse theory on Hilbert manifolds

Let us recall certain basic facts from Morse theory on Hilbert manifolds. For details, we refer the reader to the monograph of Mawhin and Willem [2].
Let M be a Riemannian manifold of class C^2 modelled on a Hilbert space H. Let $\varphi \in C^{2-0}(M,\mathbb{R})$ (in other words, let the gradient $\nabla\varphi: M \to TM$ be locally Lipschitzian). Consider the following ordinary differential equation on M

$$\dot{x}(t) = -\nabla\varphi(x(t)) \tag{1}$$

Then (1) generates a two-sided local flow π_φ on M. The *critical points of* φ, i.e. the solutions x_0 of

$$\nabla\varphi(x_0) = 0$$

are then exactly the equilibria of π_φ.

Let N be a closed subset in M. We say that φ satisfies the <u>Palais-Smale condition</u> (PS) <u>on</u> N if for every sequence $x_n \in N$ for which $\{\varphi(x_n) \mid n \in \mathbb{N}\}$ is bounded and $\nabla\varphi(x_n) \to 0$ as $n \to \infty$, it follows that $\{x_n\}$ contains a convergent subsequence. Of course, every limit point of the sequence $\{x_n\}$ is a critical point of φ.
We will now examine the relationship between the admissibility condition and the (PS)-condition. Intuitively, admissibility is an asymptotic compactness condition for solutions staying longer and longer in N.
The next result shows that under reasonable assumptions, solutions through points x_n in Int N with $\nabla\varphi(x_n) \to 0$, stay longer and longer in N, in both time direction. Thus admissibility implies the Palais-Smale condition. To keep the presentation free of differential geometric technicalities, we assume that M is an open subset of a Hilbert space H.

<u>Lemma 4.1</u>

<u>Let X be a Banach space, $U \subset X$ be open and $f: U \to X$ be locally Lipschitzian. Consider the following ordinary differential equation</u>

$$\dot{x} = f(x(t)) \ . \tag{3}$$

<u>Let π be the two-sided local flow on U generated by (3). Let $N \subset U$ be a set on which f is Lipschitzian with constant L. Let $x: [0, t_0] \to N$, $t_0 > 0$, be a solution of (3). Then for all $t \in [0, t_0]$</u>

$$\|x(t) - x(0)\| \leq \|f(x(0))\| \cdot L^{-1} \cdot e^{Lt} \ . \tag{4}$$

<u>Proof</u>:

For all $t \in [0, t_0]$ we have

$$\|x(t)-x(0)\| \leq \int_0^t \|f(x(s))-f(x(0))\| ds + \int_0^t \|f(x(0))\| ds \leq$$

$$\leq L\int_0^t \|x(s)-x(0)\| ds + \|f(x(0))\| \cdot t \ . \tag{5}$$

Let a(t) be a solution of

$$\dot{a}(t) = La(t) + b \quad a(0) = \varepsilon$$

where $b=\|f(x(0))\|$ and $\varepsilon>0$. Then integrating and using a simple argument we obtain

$$\|x(t)-x(0)\|<a(t) = \varepsilon\cdot e^{Lt}+b\cdot L^{-1}(e^{Lt}-1). \tag{6}$$

Letting $\varepsilon\to 0$ in (6) we obtain (4) and the lemma follows.

Theorem 4.2

Assume all hypotheses of Lemma 4.1. Suppose N is closed in X and π-admissible.
If $\tilde{N}$ is a subset of N such that there is a $c>0$ satisfying $\|x-y\|\geq c$ for all $x\in\tilde{N}$, $y\in\partial N$, then every sequence $\{x_n\}$ in $\tilde{N}$ with $f(x_n)\to 0$ as $n\to\infty$, contains a convergent subsequence.

Proof:

Let $x_n\in\tilde{N}$ be a sequence with $f(x_n)\to 0$. Let $s^-(x_n)=\sup\{t\geq 0\,|\,x_n\pi[-t,0]$ is defined an $\subset N\}$. Let for $t\in[0,s^-(x_n))$, $y_n(t)=x_n\pi(-t)$.
Since $\dot{y}_n(t)=-f(y_n(t))$ for $t\in[0,s^-(x_n))$, we obtain from Lemma 4.1

$$\|y_n(t)-y_n(0)\| \leq \|f(x_n)\|\cdot L^{-1}e^{Lt} \tag{7}$$

for $t\in[0,s^-(x_n))$.
(7) implies that either $s^-(x_n)=+\infty$, or else $y_n(s^-(x_n))$ is defined and $y_n(s^-(s_n))\in\partial N$.
Since $f(x_n)\to 0$ as $n\to\infty$, $s^-(x_n)<\infty$ now implies

$$c \leq \|y_n(s^-(x_n))-x_n\| \leq \|f(x_n)\|\cdot L^{-1}\cdot e^{Ls^-(x_n)}. \tag{8}$$

(8) implies that $s^-(x_n)\to\infty$ as $n\to\infty$.
Define

$$t_n = \begin{cases} n, & \text{if } s^-(x_n) = \infty \\ s^-(x_n), & \text{otherwise} \end{cases}.$$

Let $z_n=y_n(t_n)$.
It follows that $t_n\to\infty$ as $n\to\infty$, $z_n\pi[0,t_n]\subset N$ and $z_n\pi t_n=x_n$ for all n.
By admissibility of N, $\{x_n\}$ contains a convergent subsequence.
The theorem is proved.

Corollary 4.3

Suppose M is an open subset of a Hilbert space H. Let $\varphi: M \to \mathbb{R}$ be of class C^{2-0}.
Suppose that $\tilde{N}$ and N are closed in H, $\tilde{N} \subset N \subset U$, and for some $c>0$, $\|x-y\| \geq c$ for all $x \in \tilde{N}$, $y \in \partial N$. If $\nabla\varphi$ is Lipschitzian on N and N is π_φ-admissible, then φ satisfies the Palais-Smale condition on $\tilde{N}$.

We will now give an admissibility criterion for semiflows generated by parabolic equations for which sectorial operator A does not necessarily have compact resolvent.

Theorem 4.4

Let X be a Banach space and A be sectorial on X. Suppose that there is a direct sum decomposition $X = X_1 \oplus X_2$ with $A(D(A) \cap X_i) \subset X_i$, $A_i := A|X_i$, $i=1,2$, such that $\dim X_2 < \infty$ A_i is sectorial on X_i, $i=1,2$ and $\operatorname{re} \sigma(A_1) > \delta > 0$ for some $\delta > 0$.

Let $0 \leq \alpha < 1$ and U be open in X^α and $f: U \to X$ be locally Lipschitzian. Let N be closed and bounded in X^α, $N \subset U$ and $f(N) \subset C$, where C is a compact set in X. Then N is π-admissible.

Proof:

By Theorem 1.4.3 in Henry [1] there is a $C_\alpha > 0$ such that

$$\|e^{-A_1 t} u_1\|_\alpha \leq C_\alpha t^{-\alpha} e^{-\delta t} \|u_1\|_\alpha \tag{9}$$

for $u \in X_1^\alpha$.
Let $h>0$ be such that

$$k := C_\alpha h^{-\alpha} e^{-\delta h} < 1 \;. \tag{10}$$

It is easily seen that the set

$$\tilde{C} = \bigcup_{0 \leq t \leq h} e^{-A_1 t}(C)$$

is compact.
By the definition of the integral, if $y: [0,h] \to \tilde{C}$ is continuous, then $\frac{1}{h} \int_0^h y(s)\,ds \in \operatorname{Cl} \operatorname{co}(\tilde{C})$. Here, $\operatorname{co}(\tilde{C})$ is the convex hull of $\tilde{C}$. Since $\operatorname{Cl} \operatorname{co}(\tilde{C})$ is compact by Mazur's theorem, we get the existence of a compact set $D \subset X$ such that for every continuous function $u: [0,h] \to N$,

$$\int_0^h e^{-A_1 s}\cdot P_1 f(u(s))ds \in D. \tag{11}$$

Here, P_1 is the projector onto X_1 along X_2. Write $P_2=I-P_1$.
Now let β be the Kuratowski measure of noncompactness on X^α. Moreover, let $t_n\to\infty$ as $n\to\infty$, and $u_n\pi[0,t_n]\subset N$ for all n.
Write $u_n=u_n^1+u_n^2 \in X_1\oplus X_2$.
Since $\dim X_2<\infty$ and N is bounded in X^α, it follows that the sequence $P_2(u_n\pi t_n)$, $n\in\mathbb{N}$, is relatively compact. We only have to show that $P_1(u_n\pi t_n)$, $n\in\mathbb{N}$, is relatively compact.
Now, if $t_n\geq h$,

$$P_1(u_n\pi t_n)=e^{-A_1 h}P_1(u_n\pi(t_n-h))+\int_{t_n-h}^{t_n} e^{-A_1(t_n-s)}P_1(f(u_n\pi s))ds \tag{12}$$

Therefore, since $t_n\to\infty$, we get from (9), (10), (11) and (12)

$$\begin{aligned}\beta\{P_1(u_n\pi t_n)\,|\,n\in\mathbb{N}\} &= \beta\{P_1(u_n\pi t_n)\,|\,t_n\geq h\}\leq\\ &\leq \beta\{e^{-A_1 h}P_1(u_n\pi(t_n-h))\,|\,t_n\geq h\}+\beta(D)\\ &\leq k\beta\{P_1(u_n\pi(t_n-h))\,|\,t_n\geq h\}.\end{aligned} \tag{13}$$

Repeating this argument, we get for every $m\in\mathbb{N}$,

$$\beta\{P_1(u_n\pi t_n)\,|\,n\in\mathbb{N}\}\leq k^m\cdot\beta\{P_1(u_n\pi(t_n-mh))\,|\,t_n\geq mh\}\leq k^m\cdot\beta(N) \tag{14}$$

Letting $m\to\infty$, we obtain

$$\beta\{P_1(u_n\pi t_n)\,|\,n\in\mathbb{N}\} = 0$$

and the theorem follows.

Remarks:

(1) If $f_n:U\to X$ is a sequence of locally Lipschitzian mappings, N is closed and bounded in X_α, C is compact in X and $f_n(N)\subset C$ for all $n\in\mathbb{N}$, then the same proof shows that N is $\{\pi_n\}$-admissible, where $\pi_n=\pi_{f_n}$.

(2) The theorem applies in particular, to the local flow π_φ generated by equation (1), where M is open in H and $\varphi:M\to\mathbb{R}$ is a C^{2-0}-function satisfying the following assumption: $\nabla\varphi(u)=Au-f(u)$, where $A:H\to H$ is bounded, $H=H_1\oplus H_2$, $A(H_i)\subset H_i$, $A_i:=A|H_i$, $\dim H_2<\infty$, re $\sigma(A_1)>\delta>0$ for some $\delta>0$, and $f:M\to H$ is compact.

This is a situation frequently occurring in applications.
(3) There are cases in which $\varphi:M\to\mathbb{R}$ satisfies a Palais-Smale condition on a set N but N is not π_φ-admissible. For example, let H be a Hilbert space with dim $H=\infty$, and H_1, H_2 be two subspaces of H with dim $H_i=\infty$, $i=1,2$, and $H_2=H_1^\perp$. Let $u=u_1+u_2$ be the corresponding decomposition of $u\in H$.
Define $\varphi(u)=1/2\ (\|u_1\|^2-\|u_2\|^2)$. Then $\nabla\varphi(u)=Au=u_1-u_2$ for $u\in H$. Whenever $u^n=u_1^n+u_2^n$ is a sequence of elements of H with $\nabla\varphi(u^n)\to 0$ as $n\to\infty$, then, clearly, $u^n\to 0$ as $n\to\infty$. Thus the Palais-Smale condition is satisfied. However, if N is, say the closed unit ball in H, and e_n is an orthonormal sequence in H_2 with $\|e_n\|=1$ for $n\in\mathbb{N}$, then $e_n\in A^-_{\pi_\varphi}(N)$, but $\{e_n \mid n\in\mathbb{N}\}$ is not relatively compact.
We will show later on (Theorem 4.10) that cases like the one discussed in this example, are of no importance in the Morse theory, since all the critical groups (to be defined below) are zero in such a situation.
Now suppose that ξ is an isolated critical point of φ. Let $c=\varphi(\xi)$ and $\varphi^c=\{x\in M \mid \varphi(x)\leq c\}$. Let $\hat{H}_q$ be the singular homology theory with coefficients in a field F.
Then the <u>critical groups</u> $C_q(\varphi,\xi)$ of (φ,ξ) are defined to be

$$C_q(\varphi,\xi) = \hat{H}_q(\varphi^c\cap B, \varphi^c\cap B\setminus\{\xi\})\ ,\ q\in Z \qquad (15)$$

where B is any closed neighborhood of ξ. The excision property of homology implies that the critical groups are independent (up to isomorphisms) of the choice of B. Critical groups are used in the Morse theory to prove the Morse inequalities under the Palais-Smale condition. In fact we have the following

<u>Theorem 4.5</u> (see Mawhin and Willem [2])

<u>Assume the following hypotheses:</u>

(1) M <u>is a complete Riemannian manifold of class</u> C^2, <u>and</u> $\varphi:M\to\mathbb{R}$ <u>is of class</u> C^{2-0},
(2) X <u>is the closure of an open subset of</u> M. <u>Moreover,</u> X <u>is</u> π_φ<u>-positively invariant</u>.
(3) $a<b$ <u>are real numbers such that the set</u> $D=\varphi^{-1}([a,b])\cap X$ <u>contains only a finite number of critical points</u> u_j, $j=1,\dots,k$, <u>all lying in</u> Int D.
(4) <u>The numbers</u> $M_q=\sum_{j=1}^{k} C_q(\varphi,u_j)$ <u>are finite for all</u> $q\in Z$ <u>and zero for</u> q <u>sufficiently large</u>.

(5) _The Palais-Smale condition is satisfied on_ D.
Define for $d \in \mathbb{R}$, $X^d=\{x \in X \mid \varphi(x) \leq d$ _and_ $B_q = \operatorname{rank} \hat{H}_q(X^b, X^a)$, $q \in Z$.
Then, under the assumptions (1)-(5) B_q _is finite and equal zero for all_ q _large enough. Moreover, there exists a polynomial_ Q(t), _with nonnegative integer coefficients such that_

$$\sum_{n=0}^{\infty} M_n t^n = \sum_{n=0}^{\infty} B_n t^n + (1+t)Q(t) \quad . \tag{16}$$

Formula (16) is identical with formula (6) of Theorem 3.5 for the case in question, if we can prove that

$$C_q(\varphi, \xi) = \hat{H}_q(h(\pi_\varphi, \{\xi\})) \tag{17}$$

for every isolated critical point ξ of φ and

$$B_q = \hat{H}_q(h(\pi_\varphi, K)) \tag{18}$$

where K is largest π_φ-invariant set in $Cl(X^b \setminus X^a)$. (Of course, we make the somewhat stronger admissibility assumption).
We will prove formulas (17), (18) in a general abstract setting.
For the rest of this section, unless otherwise specified, let π be an arbitrary local semiflow on a metric space X. Moreover, let H_q, $q \in Z$, be an arbitrary homology or cohomology theory with coefficients in an R-module G. We assume that H_q is defined for all pairs of topological spaces and $H_q=\{0\}$ for $q<0$.
By E we denote the set of all equilibria of π. If $\varphi: X \to \mathbb{R}$ is any function and $c \in \mathbb{R}$, we write

$$\varphi^c = X^c = \{x \in X \mid \varphi(x) \leq c\}$$

$$E_c = \{x \in E \mid \varphi(x) = c\}.$$

If $\varphi: X \to \mathbb{R}$ is continuous, then φ is called a _quasi-potential of_ π if for every $x \in X \setminus E$ there is an $\varepsilon > 0$ such that the function $t \to \varphi(x\pi t)$, $t \in [0, \varepsilon)$, is strictly decreasing. Note that if φ is continuous and $\limsup_{t \to 0^+} \frac{1}{t}(\varphi(x\pi t) - \varphi(x)) < 0$ for $x \in X \setminus E$ then φ is a quasi-potential of π.
In particular, if $\varphi \in C^{2-0}(M, \mathbb{R})$, then φ is a quasi-potential of π_φ.
If φ is a quasi-potential of π, then π is easily seen to be gradient-like with respect to φ. The converse is not true. In fact, consider the one-dimensional Levin-Nohel equation

$$\dot{x}(t) = -\int_{-1}^{0} b(\theta)g(x(t+\theta))d\theta \tag{19}$$

Here, $g:\mathbb{R}\to\mathbb{R}$ is locally Lipschitzian and $b:[-1,0]\to\mathbb{R}$ is a C^2-function with $b(-1)=0, b'(\theta)\geq 0$ and $b''(\theta)\geq 0$ for $\theta\in[-1,0]$. Moreover, we assume that there is a $\theta_0\in[-1,0]$ with $b''(\theta_0)>0$.
(19) is a retarded functional differential equation which generates a local semiflow $\tilde{\pi}$ on $C=C([-1,0],\mathbb{R})$ as in Example 2 of Section 1.1. For $\psi\in C$ define

$$V(\psi) = G(\psi(0))+\frac{1}{2}\int_{-1}^{0} b'(\theta)[\int_{\theta}^{0} g(\psi(s))ds]^2 d\theta \quad . \tag{20}$$

Here, $G(x)=\int_0^x g(s)ds$.
Then

$$\dot{V}(\psi) = \lim_{t\to 0^+}\frac{1}{t}(V(\psi\tilde{\pi}t)-V(\psi)) = -\frac{1}{2}b'(-1)[\int_{-1}^{0} g(\psi(\theta))d\theta]^2$$

$$-\frac{1}{2}\int_{-1}^{0} b''(\theta)[\int_{\theta}^{0} g(\psi(s))ds]^2 d\theta \quad . \tag{21}$$

Now a simple analysis of (21) easily shows that $\tilde{\pi}$ is gradient-like with respect to V (see also Hale [1], Hale and Rybakowski [1], Kuen and Rybakowski [1]). Moreover, the equilibria of $\tilde{\pi}$ are constant functions $\psi(\theta)\equiv a$ with $g(a)=0$. Now choose g such that $g(a)=0$ for $a\in[-1,0]$. Let $\psi(\theta)=\theta$ for $\theta\in[-1,0]$. Then ψ is not a constant function so it is not an equilibrium of $\tilde{\pi}$. However, it is clear that $(\psi\tilde{\pi}t)(\theta)=0$ for all $t\geq 0$, $\theta\in[-1,0]$ with $t+\theta\geq 0$ and $(\psi\tilde{\pi}t)(\theta)=t+\theta$ if $t+\theta\leq 0$. Thus $V(\psi\tilde{\pi}t)\equiv$const from (20) and (21).
This shows that V is not a quasipotential of $\tilde{\pi}$.
For the rest of this section let φ be a quasi-potential of π.

We then have the following

Theorem 4.6

Suppose $a,b\in\mathbb{R}$, $a<b$ *are such that* $E_a=E_b=\emptyset$. *Let* $B=\{x\in X\mid a\leq\varphi(x)\leq b\}$. *Then* B *is an isolating block for* π *with*

$$B^- = \{x\in X\mid \varphi(x) = a\} \quad .$$

If π *does not explode in* B, *then there exist the following canonical isomorphisms*

$$H_q(X^b,X^a) \cong H_q(X^b/X^a,[X^a]) \cong H_q(B/B^-,[B^-]) \tag{22}$$

Proof:

By our assumptions it is easily shown that B is an isolating block and $B^{-}=\{x \in X \mid \varphi(x)=a\}$.
Define for $x \in X^{b} \setminus A^{+}(B)$

$$\tau(x) = \begin{cases} s_B(x) & \text{if } x \in B \\ 0 & \text{if } x \in X^a \end{cases}$$

By Lemma I.3.8, τ is continuous.
Set $H(x,\sigma)=x\pi(\sigma\cdot\tau(x))$, $x \in X^{b} \setminus A^{+}(B)$, $\sigma \in [0,1]$.
Using this definition of H and Proposition I.3.6 it is easily seen that the inclusion $X^a \subset X^b$ is a cofibration. Now Proposition I.10.8 implies the first isomorphism in (22). By Proposition I.6.2 there are continuous inclusion induced maps $i: B/B^{-} \to X^b/X^a$, $e: X^b/X^a \to B/B^{-}$.
It is immediate that i and e are inverse to each other. Thus i is a base-point preserving homeomorphism. Thus the second isomorphism in (22) follows.

Corollary 4.7

If the set B defined in Theorem 4.6 is strongly π-admissible and K is the largest π-invariant set in B, then $h(\pi,K)$ is defined and

$$H_q(h(\pi,K)) \cong H_q(X^b,X^a) \qquad q \in Z \tag{23}$$

i.e. *formula* (18) *holds*.
The proof is obvious.
Now, let ξ be an isolated equilibrium of π with $\xi \in E_c$. Define the general *critical groups* $C_q(\varphi,\xi)$ as

$$C_q(\varphi,\xi) = H_q(\varphi^c \cap B, \varphi^c \cap B \setminus \{\xi\}), \; q \in Z \tag{24}$$

where B is any closed neighborhood of ξ.
The excision axiom of (co)homology theory implies that $C_q(\varphi,\xi)$ is independent (up to an isomorphism) of the choice of B.
Our main result in this section is now

Theorem 4.8

Let $\xi \in E_c$ be an isolated equilibrium of π admitting a strongly π-admissible closed neighborhood N. Then the homotopy index $h(\pi,K)$, where $K=\{\xi\}$, is defined and

$$H_q(h(\pi,K)) \cong C_q(\varphi,\xi) \quad , \quad q \in \mathbb{Z} . \tag{25}$$

In other words, the critical groups of (φ,ξ) are identical with the (co)homological index of $(\varphi,\{\xi\})$.

Proof:

Since π is gradient-like with respect to φ, N is an isolating neighborhood of $K=\{\xi\}$. It follows that $h(\pi,K)$ is defined. Moreover, if $B\subset N, \xi \in B$ is any isolating block (for π), then

$$H_q(h(\pi,K)) \cong H_q(B,B^-), \quad q \in \mathbb{Z} . \tag{26}$$

We will show that there exists an isolating block $B\subset N$ with

$$H_q(B,B^-) \cong H_q(\varphi^c \cap B, \varphi^c \cap B \setminus \{\xi\}) \quad , \quad q \in \mathbb{Z} \tag{27}$$

thereby completing the proof:
By Lemma 4.9 below, there is an $\alpha<c$ and isolating block $B\subset N, \xi \in B$, with $B^- \subset \varphi^\alpha$.
Let $C=\varphi^c \cap B\setminus\{\xi\}$. Then $C\cap A^+(B)=\emptyset$. In fact suppose $x \in C\cap A^+(B)$. Then $x\pi t\to\xi$ as $t\to\infty$. Therefore $x\neq\xi$ implies $\varphi(x)>\varphi(\xi)=c$, a contradiction.
Define $s_B:B\to\mathbb{R}^+\cup\{\infty\}$ as in Lemma I.3.8 and let $D:C\times[0,1]\to C$ be defined as $D(x,t)=x\pi(t\cdot s_B(x))$.
It is easily seen that D is well-defined, and continuous .
Moreover, $B^-\subset C$ and, clearly, $s_B(x)=0$ for $x \in B^-$, so $D(x,\tau)\equiv x$ for $x \in B^-$.
Moreover, $D(x,0)\equiv x$ and $D(x,1) \in B^-$, for $x \in C$.
This proves that D is a strong deformation retraction of C onto B^-.
It follows that

$$H_q(B,B^-) \cong H_q(B,\varphi^c \cap B\setminus\{\xi\}), \quad q \in \mathbb{Z} \tag{28}$$

Now notice that if $x \in B\setminus A^+(B)$ and $\varphi(x)>c$, then $\varphi(x\pi s_B(x))\leq\alpha<c$ so that there is a unique $t(x)$ such that $\varphi(x\pi t(x))=c$. Define for $(x,\tau) \in B\times[0,1]$

$$\rho(x,\tau) = \begin{cases} x & \text{if } \varphi(x)\leq c , \\ x\pi(\tau/(1-\tau)) & \text{if } \varphi(x)>c, x \in A^+(B) ,\ 0\leq\tau<1, \\ \xi & \text{if } \varphi(x)>c, x \in A^+(B), \tau=1, \\ x\pi(\tau/(1-\tau)) & \text{if } \varphi(x)>c, x \notin A^+(B), 0\leq\tau<1,\ \tau/(1-\tau)\leq t(x) \\ x\pi t(x) & \text{if } \varphi(x)>c, x \notin A^+(B), 0\leq\tau<1 \text{ and } \tau/(1-\tau)>t(x) \\ x\pi t(x) & \text{if } \varphi(x)>c, x \notin A^+(B),\ \tau=1 . \end{cases} \tag{29}$$

ρ is well-defined, of course. Moreover, we claim that ρ is continuous.

Assume this claim for a moment. Clearly $\rho(x,\tau)\equiv x$ for $x\in\varphi^c\cap B$, $\rho(x,0)\equiv x$ and $\rho(x,1)\in\varphi^c\cap B$ for $x\in B$. Consequently ρ is a strong deformation retraction of B onto $\varphi^c\cap B$.
Thus we obtain that

$$H_q(B,\varphi^c\cap B\setminus\{\xi\}) \cong H_q(\varphi^c\cap B,\varphi^c\cap B\setminus\{\xi\}),\quad q\in Z\ . \tag{30}$$

(28) and (30) imply (27) and so the theorem is proved except for the claim.
Suppose ρ is not continuous. Then there are $\beta>0$ and sequence $(x_n,\tau_n)\in B\times[0,1]$ converging to $(x,\tau)\in B\times[0,1]$ such that

$$d(\rho(x_n,\tau_n),\ \rho(x,\tau)) \geq \beta \tag{31}$$

for all $n\in\mathbb{N}$.
We will consider several cases, each time arriving at a contradiction.

1. $\varphi(x)\leq c$:
Then $\rho(x,\tau)\equiv x$. By (31), we may assume that $\varphi(x_n)>c$ for all n. Thus $\varphi(x)=c$.
1.1 $x\in A^+(B)$:
Then, clearly, $x=\xi$ and so $x_n\to\xi$ as $n\to\infty$. Moreover, by (29), for every n, $\rho(x_n,\tau_n)=x_n\pi t_n$ for some $t_n\geq 0$ and $\varphi(x_n\pi t_n)\geq c$, or else $\rho(x_n,\tau_n)=\xi$.
By (31) we may assume the first case for all n.
We claim that $x_n\pi t_n\to\xi$ for $n\to\infty$. In fact, this is so if $\{t_n\}$ is bounded, since $\xi\pi t\equiv\xi$ for all $t\geq 0$. If $\{t_n\}$ is unbounded, then w.l.o.g. $x_n\pi t_n\to z\in A^-(B)$ and $\varphi(z)\geq c$. Let $\sigma:(-\infty,0]\to B$ be a solution with $\sigma(0)=z$. Then $\varphi(\sigma(t))\geq c$ for all $t\in\mathbb{R}^-$. Moreover, $\lim_{t\to-\infty}\sigma(t)=\xi$. Since $t\to\varphi(\sigma(t))$ is nonincreasing, it follows that $\varphi(\sigma(t))\equiv c$ for all $t\in\mathbb{R}^-$, which implies that $\sigma(t)\equiv\xi$ for all $t\in\mathbb{R}^-$, and so $z=\xi$. The claim is proved.
It follows that $\rho(x_n,\tau_n)\to\xi=\rho(x,\tau)$, a contradiction to (31).

1.2 $x\notin A^+(B)$:
We may assume that $x_n\notin A^+(B)$ for all n since $A^+(B)$ is closed. We claim that $t(x_n)\to 0$ as $n\to\infty$. In fact, let $\varepsilon>0$ be any small number. Since $x\neq\xi$, $\varphi(x)=c$, we obtain $\varphi(x\pi\varepsilon)<\varphi(x)=c$. Thus $\varphi(x_n\pi\varepsilon)<c$ for all n large enough. Hence $t(x_n)<\varepsilon$ for all such n, and the claim follows.
By (29) $\rho(x_n,\tau_n)=x_n\pi s_n$ with $0\leq s_n\leq t(x_n)$. Now the claim implies $\rho(x_n,\tau_n)\to x\pi 0=x=\rho(x,\tau)$ a contradiction to (31).

2. $\varphi(x)>c$:
Then we may assume that $\varphi(x_n)>c$ for all n.

2.1 $x \in A^+(B)$:
Then by (31) we may assume that $x_n \notin A^+(B)$ for all n.

2.1.1 $\{t(x_n)\}$ is unbounded:
Then w.l.o.g. $t(x_n)\to\infty$ as $n\to\infty$: If $\tau<1$, then $\tau_n(1-\tau_n)^{-1}\to\tau(1-\tau)^{-1}$ as $n\to\infty$, hence, in particular, $\tau_n(1-\tau_n)^{-1}<t(x_n)$ for n large enough. Then (29) implies $\rho(x_n,\tau_n)\to\rho(x,\tau)$, a contradiction. If $\tau=1$, then $\rho(x,\tau)=\xi$ and $\rho(x_n,\tau_n)=x_n\pi t_n$ with $t_n\to\infty$ and $\varphi(\rho(x_n,\tau_n))\geq c$. An argument as in case 1.1 implies $x_n\pi t_n\to\xi$, and again, a contradiction follows.

2.1.2 $\{t(x_n)\}$ is bounded:
Then, w.l.o.g. $t(x_n)\to t_0<\infty$ as $n\to\infty$. It follows that $c=\varphi(x_n\pi t(x_n))\to\varphi(x\pi t_0)$, so $\varphi(x\pi t_0)=c$. But $x\pi t_0 \in A^+(B)$, hence $x\pi t_0=\xi$. (Note that this cannot happen if π is a local (two-sided) flow). Thus $x\pi t\equiv\xi$ for all $t\geq t_0$. This and (29) clearly imply that

$$\rho(x_n,\tau_n)\to\rho(x,\tau)$$

as $n\to\infty$, a contradiction.

2.2 $x \notin A^+(B)$:
Then we may assume that $x_n \notin A^+(B)$ for all n: We claim that $t(x_n)\to t(x)$. In fact, if $M>0$, $M<t(x)$ is given, then $\varphi(x\pi M)>c$ so $\varphi(x_n\pi M)>c$ for n large so $t(x_n)>M$ for n large. Similarly, $t(x)<M$ implies $t(x_n)<M$ for n large and the claim follows. Our claim and (29) now clearly imply that $\rho(x_n,\tau_n)\to\rho(x,\tau)$ again a contradiction.
The proof is complete.

Lemma 4.9

Under the assumption of Theorem 4.8, there is an isolating block B, $B\subset N$, $\xi \in B$, and an $\alpha<c$ such that $B^-\subset\varphi^\alpha$.

Proof:

Choose $\delta>0$ such that for $0<\delta_1,\delta_2<\delta/2$, the set $B=B_{\delta_1,\delta_2}$ defined in (1) of the proof of Theorem I.5.1 is an isolating block, $\xi \in B\subset N$. Choose $0<\delta_1<\delta/2$. We will show that there are a δ_2, $0<\delta_2<\delta/2$ and $\alpha<c$ such that $B^-_{\delta_1,\delta_2}\subset\varphi^\alpha$. In fact, otherwise there is a sequence $x_n \in \tilde{U}$ with $\tilde{g}^+(x_n)=\delta_1$, $\varphi(x_n)\geq c-n^{-1}$ for $n \in \mathbb{N}$, and $\tilde{g}^-(x_n)\to 0$ as $n\to\infty$. This implies that w.l.o.g. $x_n\to z \in A^-(\tilde{N})$. Let $\sigma:\mathbb{R}^-\to\tilde{N}$ be a solution of π with $\sigma(0)=z$. By admissibility, $\sigma(t)\to\xi$ as $t\to-\infty$. Since $\tilde{g}^+$ and $\tilde{g}^-$ are continuous on $\tilde{N}$, it follows that $\tilde{g}^+(z)=\delta_1>0$, so $z\neq\xi$. Therefore since φ is a quasi-potential for π, $\varphi(z)<\varphi(\xi)=c$. However, by assumption, $\varphi(z)\geq c$, a contradiction which completes the proof.

We shall now specialize to the local flow π_φ defined by equation (1). If ξ is an isolated critical point of (i.e. an isolated equilibrium of π_φ) then the critical groups of (φ, ξ) are defined even if there is no strongly π_φ-admissible neighborhood of ξ.

However, using a result of Mawhin and Willem, we will prove that in such a case, under some additional assumption, all critical groups are zero, so they play no role in the Morse inequalities. Since all considerations are of local character, we may assume that M is an open set in a Hilbert space H, $\xi=0$ and $\varphi(\xi)=0$.

Consider the following assumption (S):

(S) U is an open neighborhood of 0 in a Hilbert space H, $\varphi: U \to \mathbb{R}$ is a C^2-function with $\varphi(0)=0$, $\nabla\varphi(0)=0$, $\nabla\varphi(u)\neq 0$ for $u\neq 0$, and $L=\varphi''(0): H \to H$ is a Fredholm operator.

We use some arguments from Mawhin and Willem [1]. Let R(L) (resp. N(L)) be the range (resp. the kernel) of L. Let $Q: H \to H$ be the orthogonal projector of H onto R(L). Write $u=v+w$ where $v=Qu$ and $w=(I-Q)u$. Then $v \in R(L)$ and $w \in N(L)$. Now the implicit function theorem implies that there is a ball $\Omega \subset H$ centered at zero and a unique C^1-map $g: \Omega \cap N(L) \to R(L)$ such that $g(0)=0$ and

$$Q\nabla\varphi(w+g(w)) = 0 \qquad \text{for } w \in \Omega \cap N(L) \tag{32}$$

Let

$$\hat{\varphi}(w) = \varphi(w+g(w)). \tag{33}$$

Then $\hat{\varphi}$ is a C^2-function (!) and

$$\nabla\hat{\varphi}(w) = (I-Q)\nabla\varphi(w+g(w)) \tag{34}$$

$$\hat{\varphi}''(w) = (I-Q)\varphi''(w+g(w))(Id+g'(w)) \tag{35}$$

for $w \in \Omega \cap N(L)$.

By the generalized Morse lemma due to Mawhin and Willem [1], there is an open neighborhood $\tilde{U} \subset U$ of 0 in H, an open neighborhood $\tilde{W} \subset \Omega \cap N(L)$ of 0 in N(L), and a homeomorphism h from $\tilde{U}$ into U such that $h(0)=0$ and

$$\varphi(h(u)) = 1/2(Lv,v)+\hat{\varphi}(w) \tag{36}$$

for $u=v+w \in \tilde{U}$.

Let $R(L)=H^+ \oplus H^-$ be the direct sum decomposition of $R(L)$ into the subspaces H^+ and H^- on which L is positive-(resp. negative-) definite, and let $v=v^+ + v^-$ be the corresponding decomposition of $v \in R(L)$.
We can now state

Theorem 4.10

Let (S) be satisfied and $\hat{\varphi}$ be as in (33). Let $m=\dim H^-$. Then the following properties hold:

1. *If* $m=\infty$, *then all critical groups of* $(\varphi,0)$ *are trivial.*
2. *If* $m<\infty$, *then* $K=\{0\}$ *has a strongly* $\pi_{\hat{\varphi}}$*-admissible isolating neighborhood,* $h(\pi_\varphi,K)$ *is defined and for all* $q \in Z$

$$C_q(\varphi,0) \cong H_q(h(\pi_\varphi,\{0\})) \cong H_{q-m}(h(\pi_{\hat{\varphi}},\{0\})) \cong C_q(\hat{\varphi},0). \qquad (37)$$

Here, $\pi_{\hat{\varphi}}$ *is the local semiflow on* $\Omega \cap N(L)$ *generated by* $\dot{w}=-\nabla\hat{\varphi}(w)$.

Proof:

Let B be any closed bounded neighborhood of 0 in H, $B \subset \tilde{U}$.
Then

$$C_q(\varphi,0) \cong H_q(\varphi^0 \cap h(B), \varphi^0 \cap h(B) \setminus \{0\}) \cong H_q(\psi^0 \cap B, \psi^0 \cap B \setminus \{0\}) \cong C_q(\psi,0) \quad (38)$$

where

$$\psi(u) = 1/2(Lv,v) + \hat{\varphi}(w). \qquad (39)$$

Consider the ODE

$$\dot{u} = -\nabla\psi(u) . \qquad (40)$$

(40) is uncoupled and has the form

$$\begin{aligned} \dot{v}^+ &= -Lv^+ \\ \dot{v}^- &= -Lv^- \\ \dot{w} &= -\nabla\hat{\varphi}(w) . \end{aligned} \qquad (41)$$

Let us prove that 0 is the only critical point of ψ in $\tilde{U}$. In fact, $\nabla\psi(0)=0$ of course. If $\nabla\psi(u)=Lv+\nabla\hat{\varphi}(w)=0$, then $v=0$ and $\nabla\hat{\varphi}(w)=0$. Now (32) and (34) imply that $\nabla\varphi(w+g(w))=0$, so $w+g(w)=0$, i.e. $w=0$.
The claim follows.

If $m<\infty$, then by remark (2) after the proof of Theorem 4.4 $\{0\}$ has a strongly π_φ-admissible isolating neighborhood, so $h(\pi_\varphi,\{0\})$ is defined. Here we have used the fact that $\dim N(L)<\infty$. Since π_ψ is a product of local flows, we obtain

$$h(\pi_\psi,\{0\}) = \textstyle\sum^m \wedge h(\pi_{\hat{\varphi}},\{0\}) \ . \tag{42}$$

Hence, using arguments from the proof of Proposition II.6.4 we obtain

$$H_q(h(\pi_\psi,\{0\})) = H_{q-m}(h(\pi_{\hat{\varphi}},\{0\})) \tag{43}$$

(38), (42), (43) and Theorem 4.8 imply formula (37).
Now we assume that $m=\infty$. By our assumption $B\cap N(L)$ is an isolating neighborhood of $\{0\}$ with respect to $\pi_{\hat{\varphi}}$. Moreover, since $\dim N(L)<\infty$, $B\cap N(L)$ is trivially strongly $\pi_{\hat{\varphi}}$-admissible.
Let $\varepsilon>0$ be arbitrary and set

$$B_1 = \{v^+ \in H^+ \mid (Lv^+,v^+)\leq\varepsilon\} \tag{44}$$

$$B_2 = \{v^- \in H^- \mid -(Lv^-,v^-)\leq\varepsilon\} \tag{45}$$

Since $\hat{\varphi}(0)=0$, there is a ball W in $N(L)$ at 0 with radius $<\varepsilon$ such that

$$|\hat{\varphi}(w)| < \varepsilon/2 \quad \text{for } w \in W \ . \tag{46}$$

Choose $\varepsilon>0$ so small that $B_1\oplus B_2\oplus W\subset\tilde{U}$.
By Lemma 4.9, there is an isolating block $B_3\subset W$, $0\in B_3$, with respect to $\pi_{\hat{\varphi}}$ such that $B_3^-\subset\hat{\varphi}^{-\delta}$ for some $\delta>0$. Choose such a block B_3 and let $B=B_1\oplus B_2\oplus B_3$. Define for $\tau\in[0,1]$, $u=v^++v^-+w\in B$
$D_1(u,\tau)=(1-\tau)v^++v^-+w\in B$.
D_1 is a continuous map from $B\times[0,1]$ to B. Moreover, $D_1((\psi^0\cap B)\times[0,1])\subset\psi^0\cap B$. Let $\tau\in[0,1]$ and $u\in\psi^0\cap B\setminus\{0\}$. Then $D_1(u,\tau)\neq0$ if $\tau<1$, of course. Suppose $D_1(u,1)=0$. Then $v^-+w=0$ so $u=v^+\neq0$. But then $\psi(u)>0$, a contradiction.
Thus using D_1, we see that $\psi^0\cap(B_2\oplus B_3)$ is a strong deformation retract of $\psi^0\cap B$ and $\psi^0\cap(B_2\oplus B_3)\setminus\{0\}$ is a strong deformation retract of $\psi^0\cap B\setminus\{0\}$. Thus

$$C_q(\psi,0)\cong H_q(\psi^0\cap(B_2\oplus B_3),\psi^0\cap(B_2\oplus B_3)\setminus\{0\}) . \tag{47}$$

Define for $u=v^-+w\in B_2\oplus B_3$

$s^+(u)=\sup\{t \mid u\pi_\psi[0,t] \text{ is defined and } u\pi_\psi[0,t]\subset B_2\oplus B_3\}$.
If $u\notin A^+(B_2\oplus B_3)$, then $s^+(u)<\infty$. Moreover, as in Lemma I.3.8, s^+ is seen to be continuous on $B_2\oplus B_3\setminus A^+(B_2\oplus B_3)$.
Let us show that $\psi^0\cap(B_2\oplus B_3)\setminus\{0\}\cap A^+(B_2\oplus B_3)=\emptyset$. In fact, let $u=v^-+w\in\psi^0\cap(B_2\oplus B_3)\setminus\{0\}$.
If $v^-\neq 0$, then from (41), $u\notin A^+(B_2\oplus B_3)$. Suppose $v^-=0$. Then $w\neq 0$ and $\hat{\varphi}(w)\leq 0$. Since $\hat{\varphi}$ is a potential of $\pi_{\hat{\varphi}}$, it follows that the solution of $\pi_{\hat{\varphi}}$ through $x=w$ must leave B_3. The claim is proved.
Define for $\tau\in[0,1]$, $u\in\psi^0\cap(B_2\oplus B_3)\setminus\{0\}$, $D_2(u,\tau)=u\pi_\psi(\tau\cdot s^+(u))\in B_2\oplus B_3$.
Since ψ decreases along solutions of π_ψ, $D_2(u,\tau)\in\psi^0\setminus\{0\}$.
D_2 defines a strong deformation retraction onto
$\psi^0\cap(\partial B_2\oplus B_3\cup B_2\oplus B_3^-)=\partial B_2\oplus B_3\cup B_2\oplus B_3^-$.
The last equality is a consequence of our choice of ε,δ and B_3.
Now (47) implies that

$$C_q(\psi,0)\cong H_q(\psi^0\cap(B_2\oplus B_3),\ \partial B_2\oplus B_3\cup B_2\oplus B_3^-) \quad . \tag{48}$$

Let $\rho:B_3\times[0,1]\to B_3$ be defined as in (29) with φ,ξ,π,B being replaced by $\hat{\varphi},0,\pi_{\hat{\varphi}},B_3$ respectively.
For $\tau\in[0,1]$, $u=v^-+w\in B_2+B_3$ define

$$D_3(u,\tau) = v^-+\rho(\tau,w) \quad .$$

From (29) we see that $\hat{\varphi}(\rho(w,\tau))\leq\hat{\varphi}(w)$. Hence, if $u\in\psi^0$, then $\psi(D_3(u,\tau))=\frac{1}{2}(Lv^-,v^-)+\hat{\varphi}(\rho(w,\tau))\leq\psi(u)\leq 0$.
Therefore $D_3:(\psi^0\cap(B_2\oplus B_3))\times[0,1]\to\psi^0\cap(B_2\oplus B_3)$ is well-defined and continuous.
D_3 is a strong deformation retraction onto $\psi^0\cap(B_2\oplus(\hat{\varphi}^0\cap B_3))=B_2\oplus(\hat{\varphi}^0\cap B_3)$.
Moreover, the restriction of D_3 to $(\partial B_2\oplus B_3\cup B_2\oplus B_3^-)\times[0,1]$ is easily seen to be a strong deformation retraction onto $\partial B_2\oplus(\hat{\varphi}^0\cap B_3)\cup B_2\oplus B_3^-$.
Thus

$$C_q(\psi,0)\cong H_q(B_2\oplus(\hat{\varphi}^0\cap B_3),\ \partial B_2\oplus(\hat{\varphi}^0\cap B_3)\cup B_2\oplus B_3^-) \quad . \tag{49}$$

On H^- define the norm $|v^-|=-\varepsilon^{-1}.(Lv^-,v^-)$. $|v^-|$ is equivalent to the scalar product norm on H^-. Since H^- is infinite-dimensional, a well-known result of Dugundji implies that there is a strong deformation retraction D_4 of the closed unit ball $\tilde{\Omega}$ in $(H^-,|\ |)$ onto $\partial\tilde{\Omega}$ (see

Deimling [1], page 66 for an easy proof that there is a retraction $r:\tilde{\Omega}\to\partial\tilde{\Omega}$ and define the strong deformation retraction D_4 by $D_4(x,\tau)=(1-\tau)x+\tau r(x))$.
However, $\tilde{\Omega}=B_2$, so there is a strong deformation retraction D_4 of B_2 onto ∂B_2. Define $D_5(v^-+w,\tau)=D_4(v^-,\tau)+w$.
Using D_5 and arguments as above, we get from (49)

$$C_q(\psi,0) \cong H_q(\partial B_2\oplus(\hat{\varphi}^0\cap B_3),\partial B_2\oplus(\hat{\varphi}^0\cap B_3)) = \{0\} \tag{50}$$

Now (38) and (50) complete the proof that $C_q(\varphi,0)=\{0\}$ for all $q\in\mathbb{Z}$.

3.5 Continuation of the Categorial Morse index along paths

In this section we will examine in some greater detail the problem of continuing the index along S-continuous maps $\alpha:=\Lambda\to S$.
By imposing a little extra hypothesis on α (the strict S-continuity) which is satisfied in all applications of the index considered, say, in Chapter II of this book, we will be able to "continue" the whole categorial Morse index along paths $\alpha:=[0,1]\to S$. We will see that two such paths which are "S-homotopic" to each other, give rise to the same continuation.
Although no specific applications of this generalized continuation property are given, the results obtained help us to gain some deeper understanding of what remains invariant in the structure of the neighborhood of an invariant set as the semiflow is changed continuously.
In this section unless otherwise specified X is a fixed metric space and $S:=S(X)$.
We first prove the following

Proposition 5.1

Let $\alpha:\Lambda\to S$ be S-continuous.
For $\tilde{\Lambda}\subset\Lambda$ define

$$D = D(\tilde{\Lambda}) = \{(t,(\lambda,x))\in\mathbb{R}^+\times(\tilde{\Lambda}\times X) \mid x\pi_\lambda t \text{ is defined}\}$$

For $(t,(\lambda,x))\in D$ define $(\lambda,x)\pi(\tilde{\Lambda})t:=(\lambda,x\pi_\lambda t)$. Then D is open in $\tilde{\Lambda}\times X$ and $\pi(\tilde{\Lambda})$ is a local semiflow on $\tilde{\Lambda}\times X$.
Now assume that $\tilde{\Lambda}$ is compact. Then $K(\tilde{\Lambda}):=\bigcup_{\lambda\in\tilde{\Lambda}}\{\lambda\}\times K_\lambda$ is a compact $\pi(\tilde{\Lambda})$-invariant set. Furthermore, if $N\subset X$ is an isolating neighborhood of K_λ, relative to π_λ, for $\lambda\in\tilde{\Lambda}$, then $\tilde{\Lambda}\times N$ is an isolating neighborhood of $K(\tilde{\Lambda})$, relative to $\pi(\tilde{\Lambda})$.

Finally, if $N \subset X$ is closed and strongly $\{\pi_{\lambda_n}\}$-admissible for every sequence $\{\lambda_n\} \subset \tilde{\Lambda}$, then $\tilde{\Lambda} \times N$ is strongly $\pi(\tilde{\Lambda})$-admissible.

Proof:

Suppose that D is not open. Then there exists $(t_0,(\lambda_0,x_0)) \in D$ and a sequence $(t_n,(\lambda_n,x_n))$ converging to $(t_0,(\lambda_0,x_0))$ such that $(t_n,(\lambda_n,x_n)) \notin D$ for all $n \in \mathbb{N}$. However, by the S-continuity of α, $\pi_{\lambda_n} \to \pi_{\lambda_0}$ as $n \to \infty$, an obvious contradiction.
Hence, indeed, D is open. $\pi(\tilde{\Lambda})$ is clearly a local semiflow on $\tilde{\Lambda} \times X$.
Now let $\tilde{\Lambda}$ be compact. $K(\tilde{\Lambda})$ is clearly $\pi(\tilde{\Lambda})$-invariant and we only have to show it is compact in $\tilde{\Lambda} \times X$. Let $(\lambda_n,x_n) \in K(\tilde{\Lambda})$, $n \in \mathbb{N}$. Since $\tilde{\Lambda}$ is compact, we may assume $\lambda_n \to \lambda_0$ as $n \to \infty$. Since α is S-continuous, there exists a closed set $\tilde{N} \subset X$ and an $n_0 \in \mathbb{N}$ such that for all $n \geq n_0$ and $n=0$: $\tilde{N}$ is a strongly π_n-admissible isolating neighborhood of K_{λ_n} relative to π_n, where $\pi_n := \pi_{\lambda_n}$. Moreover, $\pi_n \to \pi_0$ as $n \to \infty$, and $\tilde{N}$ is $\{\pi_{n_m}\}$-admissible for every subsequence π_{n_m} of $\{\pi_n\}$. In particular $K_{\lambda_n} = A_{\pi_n}(\tilde{N})$ for $n \geq n_0$ and $n=0$. Let $y_n \in K_{\lambda_n}$, $t_n \to \infty$ be such that $y_n \pi_n t_n = x_n$. By Theorem I.4.5 we may therefore assume that $\{x_n\}$ is convergent to some $x_0 \in A^-_{\pi_0}(\tilde{N})$. By the same theorem, $x_0 \in A^+_{\pi_0}(\tilde{N})$. Thus $x_0 \in K_{\lambda_0}$, proving that $K(\tilde{\Lambda})$ is compact.
If $N \subset X$ is an isolating neighborhood of K_λ (relative to π_λ) for $\lambda \in \tilde{\Lambda}$, then $\tilde{\Lambda} \times N$ is closed in $\tilde{\Lambda} \times X$, $\tilde{\Lambda} \times \text{Int } N$ is open in $\tilde{\Lambda} \times X$, $K(\tilde{\Lambda}) \subset \tilde{\Lambda} \times \text{Int } N$ and $K(\tilde{\Lambda})$ is the largest $\pi(\tilde{\Lambda})$-invariant set in $\tilde{\Lambda} \times N$.
Finally suppose that $N \subset X$ is closed and strongly $\{\pi_{\lambda_n}\}$-admissible for every sequence $\{\lambda_n\} \subset \tilde{\Lambda}$.
To show that $\tilde{\Lambda} \times N$ is strongly $\pi(\tilde{\Lambda})$-admissible, notice, first, that π_λ does not explode in N for $\lambda \in \tilde{\Lambda}$ and so $\pi(\tilde{\Lambda})$ does not explode in $\tilde{\Lambda} \times N$.
Now let $(\lambda_n,x_n)\pi(\tilde{\Lambda})[0,t_n] \subset \tilde{\Lambda} \times N$, $n \in \mathbb{N}$, $t_n \to \infty$.
Since $\tilde{\Lambda}$ is compact, we may assume that $\lambda_n \to \lambda_0$ as $n \to \infty$. Moreover, $x_n \pi_{\lambda_n}[0,t_n] \subset N$ for all $n \to \infty$. Thus we may assume that $\{x_n \pi_{\lambda_n} t_n\}$ is convergent, so $\{(\lambda_n,x_n)\pi(\tilde{\Lambda})t_n\}$ is convergent.
The proposition is proved.

We will now state a technical "continuation" theorem, which is an extension of Theorem I.12.3:

Theorem 5.2

Let $\alpha:\Lambda\to S$ be S-continuous and $\lambda_0\in\Lambda$. Let $\tilde{N}\subset X$ be closed and $K_{\lambda_0}\subset\mathrm{Int}\,\tilde{N}$. Then there is a neighborhood W of λ_0 in Λ, two isolating neighborhoods N' and N of K_{λ_0} (relative to π_{λ_0}) and two index pairs $\langle N_1',N_2'\rangle$, $\langle N_1,N_2\rangle$ in N' and N, respectively, (relative to π_{λ_0}) such that for every compact set $\tilde{\Lambda}\subset W$ the following properties are satisfied:

(1) $N'\subset\mathrm{Int}\,N\subset\mathrm{Int}\,\tilde{N}$, $N_1'=N'$, $N_1=N$.

(2) *$\tilde{\Lambda}\times N'$ and $\tilde{\Lambda}\times N$ are isolating neighborhoods of $K(\tilde{\Lambda})$ (relative to $\pi(\tilde{\Lambda})$).*

(3) *there are two index pairs*

$$\langle N'_{1,\tilde{\Lambda}},\ N'_{2,\tilde{\Lambda}}\rangle \quad \text{and} \quad \langle N_{1,\tilde{\Lambda}},\ N_{2,\tilde{\Lambda}}\rangle$$

in $\tilde{\Lambda}\times N'$ and $\tilde{\Lambda}\times N$, respectively (relative to $\pi(\tilde{\Lambda})$) such that

$$N_{i,\tilde{\Lambda}}\subset\tilde{\Lambda}\times N_i'\subset N_{i,\tilde{\Lambda}}\subset\tilde{\Lambda}\times N_i,\quad i=1,2.$$

The proof of Theorem 5.2 parallels that for Theorem I.12.3:

First assume that $K_{\lambda_0}=\emptyset$. Then, using the S-continuity of α, it is easily shown that $K_\lambda=\emptyset$ for all λ in a neighborhood W of λ_0 in Λ. (cf. the proof of Theorem I.12.3).
Therefore $K(\tilde{\Lambda})=\emptyset$ for $\tilde{\Lambda}\subset W$, and we may set $N=N'=N_1=N_1'=N_{1,\tilde{\Lambda}}=N_{2,\tilde{\Lambda}}=N'_{1,\tilde{\Lambda}}=N'_{2,\tilde{\Lambda}}=\emptyset$. Then properties (1)-(3) hold trivially.
Therefore we can assume, w.l.o.g. that $K_{\lambda_0}\neq\emptyset$.

We need a few lemmas:

Lemma 5.3

Let $V,U,\tilde{U}$ be open in X, and such that $\mathrm{Cl}\,V\subset U$, $N:=\mathrm{Cl}\,U\subset\tilde{U}$. Let $\tilde{\Lambda}\subset\Lambda$ be compact and assume $K_\lambda\subset V$ for $\lambda\in\tilde{\Lambda}$. Define for $\lambda\in\Lambda$, $x\in\tilde{U}$:

$t_\lambda^+(x)=\sup\{t\geq 0\mid x\pi_\lambda[0,t]$ *is defined and* $\subset\tilde{U}\}$.

Now define for arbitrary $M>0$,

$N_{1,\tilde{\Lambda}}=(\tilde{\Lambda}\times N)\cap\mathrm{Cl}_{\tilde{\Lambda}\times X}\{(\lambda,y)\in\tilde{\Lambda}\times X\mid$ *there is* $x\in V$ *and* $t\in\mathbb{R}^+$ *such that* $x\pi_\lambda[0,t]$ *is defined,* $\subset\tilde{U}$ *and* $x\pi_\lambda t=y\}$

$N_{2,\tilde{\Lambda}}=N_{1,\tilde{\Lambda}}\cap\{(\lambda,x)\in\tilde{\Lambda}\times\tilde{U}\mid t_\lambda^+(x)\leq M\}$.

Then the following properties hold:

(1) $N_{1,\tilde\Lambda}$ and $N_{2,\tilde\Lambda}$ are closed and $(\Lambda\times N)$-positively invariant subsets of $\tilde\Lambda\times N$ (relative to $\pi(\tilde\Lambda)$).

(2) $K(\tilde\Lambda)\subset(\tilde\Lambda\times V)\cap\{(\lambda,x)\in\tilde\Lambda\times U \mid t^+_\lambda(x)>M\}\subset N_{1,\tilde\Lambda}\setminus N_{2,\tilde\Lambda}$.

(3) $\tilde\Lambda\times V\subset N_{1,\tilde\Lambda}$.

Proof:

If $(\lambda,x)\in\tilde\Lambda\times V$ then let $y=x$, $t=0$, and this shows that $(\lambda,x)\in N_{1,\tilde\Lambda}$, proving (3).

If $(\lambda,x)\in K(\tilde\Lambda)$ then $x\in K_\lambda\subset V$ and $t^+_\lambda(x)=\infty$ so (2) is satisfied.

Obviously $N_{1,\tilde\Lambda}$ is closed in $\tilde\Lambda\times X$. Let $(\lambda_n,x_n)\in N_{2,\tilde\Lambda}$, $(\lambda_n,x_n)\to(\lambda_0,x_0)\in\tilde\Lambda\times X$. Then $(\lambda_0,x_0)\in\tilde\Lambda\times N\subset\tilde\Lambda\times\tilde U$. If $t^+_{\lambda_0}(x_0)>M$, then $x_0\pi_{\lambda_0}[0,M]$ is defined and $\subset\tilde U$. Since $\pi_{\lambda_n}\to\pi_{\lambda_0}$ by S-continuity, we get that for all n large enough, $x_n\pi_{\lambda_n}[0,M]$ is defined and $\subset\tilde U$. Hence $t^+_{\lambda_n}(x_n)>M$, a contradiction. Thus $t^+_{\lambda_0}(x_0)\le M$, and therefore $(\lambda_0,x_0)\subset N_{2,\tilde\Lambda}$.

This proves that $N_{2,\tilde\Lambda}$ is closed.

Now suppose that $(\lambda,x)\in N_{1,\tilde\Lambda}$ and $(\lambda,x)\pi(\tilde\Lambda)[0,s]\subset\tilde\Lambda\times N$. Then there are $(\lambda_n,y_n)\to(\lambda,x)$ as $n\to\infty$, x_n and $t_n\ge 0$ such that $x_n\pi_{\lambda_n}[0,t]\subset\tilde U$ and $x_n\pi_{\lambda_n}t_n=y_n$. Since $N\subset\tilde U$, we have $x\pi_\lambda[0,s]\subset\tilde U$. Hence by the fact that $\pi_{\lambda_n}\to\pi_\lambda$, we obtain $y_n\pi_{\lambda_n}[0,s]\subset\tilde U$ for all n large enough. Hence $x_n\pi_{\lambda_n}[0,t_n+s]\subset\tilde U$ and therefore $x\pi[0,s]\subset N_{1,\tilde\Lambda}$. This proves that $N_{1,\tilde\Lambda}$ is $\tilde\Lambda\times N$-positively invariant for $\pi(\tilde\Lambda)$. Since t^+_λ does not increase along trajectories of π_λ, it also follows that $N_{2,\tilde\Lambda}$ is $\tilde\Lambda\times N$-positively invariant.

The lemma is proved.

Now using the definition of S-continuity and Theorem I.4.5 part 2.2, and taking $\tilde N$ smaller if necessary, we may assume that:

a) $\tilde N$ is an isolating block relative to π_{λ_0}. Moreover, there is an open set $\tilde U$ with $K_{\lambda_0}\subset\tilde U$ and Cl $\tilde U=\tilde N$.

b) There is a neighborhood $\tilde W$ of λ_0 in Λ such that for every $\lambda\in\tilde W$, $\tilde N$ is a strongly π_λ-admissible isolating neighborhood of K_λ relative to π_λ and whenever $\lambda_n\to\lambda_0$, then $\tilde N$ is $\{\pi_{\lambda_n}\}$-admissible.

Define g^- as in Proposition I.5.2 with N (resp. U) replaced by $\tilde N$ (resp. $\tilde U$) and π replaced by π_{λ_0} in that proposition. Let $t^+_\lambda(x)$ be defined as in Lemma 5.3 for $x\in\tilde U$. Finally set for $a,b>0$

$$V(a,b) = \{x \in \tilde{U} \mid g^-(x)<a,\ t_\lambda^+(x)>b\} \tag{1}$$

Then we have the following lemma:

Lemma 5.4

(1) Whenever $x_n \to x$ in $\tilde{U}$ and $\lambda_n \to \lambda_0$ in Λ, then $t_{\lambda_n}^+(x_n) \to t_{\lambda_0}^+(x)$.

(2) There exist $a_0, b_0 > 0$, such that

$$N := \mathrm{Cl}\ V(a_0,b_0) \subset \tilde{U} \tag{2}$$

(3) Choose a_0, b_0 and N as in (2). Then for every $M>b_0$ there is an $\varepsilon_0=\varepsilon_0(M)$, $0<\varepsilon_0<a_0$, such that for every ε, $0<\varepsilon<\varepsilon_0$ there is a neighborhood $W_0=W_0(\varepsilon,M)$ of λ_0 in Λ such that for every compact set $\tilde{\Lambda}\subset W_0$ properties (3.1), (3.2) below hold:

(3.1) $\mathrm{Cl}\ V(\varepsilon,M) \subset U$, $K_\lambda \subset V(\varepsilon,M)$ for $\lambda \in W_0$.

(3.2) If the sets $N_{1,\tilde{\Lambda}}=N_{1,\tilde{\Lambda}}(\varepsilon,M)$ and $N_{2,\tilde{\Lambda}}=N_{2,\tilde{\Lambda}}(\varepsilon,M)$ are defined as in Lemma 5.3 with V being replaced by $V(\varepsilon,M)$, then $(N_{1,\tilde{\Lambda}}, N_{2,\tilde{\Lambda}})$ is an index pair in $\tilde{\Lambda}\times N$, relative to $\pi(\tilde{\Lambda})$.

Proof:

Parts (1) and (2) are proved exactly as in the proof of Lemma I.12.6. Now let $M>b_0$, $\varepsilon<a_0$. Then again as in the proof of Lemma I.12.6, $\mathrm{Cl}\ V(\varepsilon,M)\subset U$. We claim that there is a neighborhood $W=W(\varepsilon,M)$ of λ_0 in Λ such that $K_\lambda \subset V(\varepsilon,M)$ for $\lambda \in W$. In fact, otherwise, there is a sequence $\lambda_n \to \lambda_0$, $\lambda_n \in \tilde{W}$, with $K_{\lambda_n} \not\subset V(\varepsilon,M)$. Since $K_\lambda = A_{\pi_\lambda}(\tilde{N})$ for $\lambda \in \tilde{W}$, Theorem I.4.5 implies a contradiction.

By Lemma 5.3 the sets $\tilde{\Lambda}\times N$, $N_{1,\tilde{\Lambda}}(\varepsilon,M)$ and $N_{2,\tilde{\Lambda}}(\varepsilon,M)$ satisfy the conditions of Definition I.7.1 of index pairs, except possibly for condition (3) in that definition. Therefore, if Lemma 5.4 is not true, then there is an $M>b_0$, a sequence $\varepsilon_n \to 0$ and a sequence $\tilde{\Lambda}_n$ of compact sets with $d(\lambda_0,\tilde{\Lambda}_n)\to 0$ as $n\to\infty$, such that

$N_{1,\tilde{\Lambda}_n}(\varepsilon_n,M) \cap \partial(\tilde{\Lambda}_n\times N) \setminus N_{2,\tilde{\Lambda}_n}(\varepsilon_n,M) \neq \emptyset$.

Here, $\partial(\tilde{\Lambda}\times N)$ is the boundary relative to $\tilde{\Lambda}\times X$. Choose $(\lambda_n,y_n) \in N_{1,\tilde{\Lambda}_n}(\varepsilon_n,M)\cap\partial(\tilde{\Lambda}_n\times N)\setminus N_{2,\tilde{\Lambda}_n}(\varepsilon_n,M)$. Then there are

$(\tilde{\lambda}_n,\tilde{y}_n) \in \tilde{\Lambda}_n\times X$, $x_n \in V(\varepsilon_n,M)$, $t_n \geq 0$ such that $d((\tilde{\lambda}_n,\tilde{y}_n),(\lambda_n,y_n))<2^{-n}$,

$x_n \pi_{\tilde{\lambda}_n}[0,t_n]$ is defined, $\subset \tilde{U}$ and $x_n \pi_{\tilde{\lambda}_n} t_n = \tilde{y}_n$.

We have $\tilde{\lambda}_n \to \lambda_0$ and $\lambda_n \to \lambda_0$ for $n\to\infty$. Let $\tilde{\Lambda}=\{\lambda_0\}$.
As in the proof of Lemma I.12.6 we may assume that $x_n \to x_0 \in A^-_{\pi_{\lambda_0}}(\tilde{N})$, and $\tilde{y}_n = x_n \pi_{\tilde{\lambda}_n} t_n \to y_0 \in A^-_{\pi_{\lambda_0}}(\tilde{N})$. It follows that $(\lambda_0, y_0) \in \partial(\tilde{\Lambda}\times N)$, so $y_0 \in \partial N$. Moreover, $y_0 \in A^-_{\pi_{\lambda_0}}(\tilde{N})$ implies $g^-(y_0)=0$. By our assumptions, $t^+_{\lambda_n}(y_n)>M$ for all n. Since $y_n \to y_0$, we get from part (1) of this lemma that $t^+_{\lambda_0}(y_0) \geq M > b_0$.
It follows that $y_0 \in U \cap \partial N = \emptyset$, a contradiction which proves the lemma.

Lemma 5.5

For every $M>0$ there is an $\varepsilon_1=\varepsilon_1(M)>0$ and a neighborhood $W_1=W_1(M)$ of λ_0 in Λ such that for all ε, $0<\varepsilon\leq\varepsilon_1$ and all compact sets $\tilde{\Lambda}\subset W_1$:

$$C_{\tilde{\Lambda}}(\varepsilon) \subset D_{\tilde{\Lambda}}(\varepsilon) \subset \hat{C}_{\tilde{\Lambda}}(\varepsilon) \subset \hat{D}_{\tilde{\Lambda}}(\varepsilon)$$

where

$$\begin{aligned}
C_{\tilde{\Lambda}}(\varepsilon) &= \{(\lambda,x) \in \tilde{\Lambda}\times\tilde{U} \mid t^+_\lambda(x) \leq 2M\} \cap (\tilde{\Lambda}\times \mathrm{Cl}\ V(\varepsilon,M)) \\
D_{\tilde{\Lambda}}(\varepsilon) &= \tilde{\Lambda}\times(\{x \in \tilde{U} \mid t^+_{\lambda_0}(x) \leq 3M\} \cap \mathrm{Cl}\ V(\varepsilon,M)) \\
\hat{C}_{\tilde{\Lambda}}(\varepsilon) &= \{(\lambda,x) \in \tilde{\Lambda}\times\tilde{\tilde{U}} \mid t^+_\lambda(x) \leq 4M\} \cap N_{1,\tilde{\Lambda}}(\varepsilon,M) \\
\hat{D}_{\tilde{\Lambda}}(\varepsilon) &= \tilde{\Lambda}\times(\{x \in \tilde{U} \mid t^+_{\lambda_0}(x) \leq 5M\} \cap N) .
\end{aligned}$$

Proof:

If the lemma is not true, then, taking subsequences if necessary, we can assume that there is an $M>0$ and sequences $\tilde{\Lambda}_n \to \lambda_0$, $\varepsilon_n \to 0$ and $x_n \in X$ such that

$(\lambda_n, x_n) \in C_{\tilde{\Lambda}_n}(\varepsilon_n) \setminus D_{\tilde{\Lambda}_n}(\varepsilon_n)$ for all n, or

$(\lambda_n, x_n) \in D_{\tilde{\Lambda}_n}(\varepsilon_n) \setminus \hat{C}_{\tilde{\Lambda}_n}(\varepsilon_n)$ for all n, or

$(\lambda_n, x_n) \in \hat{C}_{\tilde{\Lambda}_n}(\varepsilon_n) \setminus \hat{D}_{\tilde{\Lambda}_n}(\varepsilon_n)$ for all n.

If, for instance, the third alternative holds, then

$$t^+_{\lambda_n}(x_n) \leq 4M \tag{3}$$

$$(\lambda_n, x_n) \in N_{1,\tilde{\Lambda}_n}(\varepsilon_n, M) \tag{4}$$

$$t^+_{\lambda_0}(x_n) > 5M \tag{5}$$

for all n.

(4) implies, as in the proof of Lemma 5.4 that $\{x_n\}$ contains a convergent subsequence. Without loss of generality, $x_n \to x_0 \in A^-_{\pi_{\lambda_0}}(\tilde{N})$.

However, by (1) of Lemma 5.4, we obtain from (3)

$$t^+_{\lambda_0}(x_0) \leq 4M$$

while (5) implies

$$t^+_{\lambda_0}(x_0) \geq 5M .$$

This contradiction implies that the third alternative cannot hold. The remaining alternatives are ruled out in the same way.

We will now complete the proof of Theorem 5.2:

Choose $a_0, b_0 > 0$ such that Cl $V(a_0,b_0) \subset \tilde{U}$. This is possible by Lemma 5.4. Let $U = V(a_0,b_0)$ and $N = \text{Cl } U$. Fix an $M > b_0$. Choose an $\varepsilon_1 = \varepsilon_1(M) > 0$, $\varepsilon_1 < a_0$, and a neighborhood $W_1 = W_1(M)$ of λ_0 such that the conclusions of Lemma 5.5 hold for $\varepsilon \leq \varepsilon_1$ and compact sets $\tilde{\Lambda} \subset W_1$.

Choose an $\varepsilon_0 \leq \varepsilon_1$, $\varepsilon_0 = \varepsilon_0(M)$ such that (3) of Lemma 5.4 holds for $\varepsilon \leq \varepsilon_0$ and neighborhoods $W_0 = W_0(\varepsilon, M)$ of λ_0.

Let $\varepsilon = \varepsilon_0$ and $W_2 = W_0(\varepsilon_0, M) \cap W_1(M)$.
Then for every compact $\tilde{\Lambda} \subset W_2$

$$\langle N_{1,\tilde{\Lambda}}(\varepsilon_0, M),\ N_{2,\tilde{\Lambda}}(\varepsilon_0, M) \rangle$$

is an index pair in $\tilde{\Lambda} \times N$, rel. to $\pi(\tilde{\Lambda})$ and $\tilde{\Lambda} \times N$ is an isolating neighborhood of $K(\tilde{\Lambda})$.
Set $N_{1,\tilde{\Lambda}} := N_{1,\tilde{\Lambda}}(\varepsilon_0, M)$ and

$$N_{2,\tilde{\Lambda}} = N_{1,\tilde{\Lambda}} \cap \{(\lambda,x) \in \tilde{\Lambda} \times \tilde{U} \mid t^+_\lambda(x) \leq 4M\} \tag{6}$$

Then $N_{2,\tilde{\Lambda}} \supset N_{2,\tilde{\Lambda}}(\varepsilon_0, M)$ and so $\langle N_{1,\tilde{\Lambda}}, N_{2,\tilde{\Lambda}} \rangle$ is an index pair in $\tilde{\Lambda} \times N$, relative to $\pi(\tilde{\Lambda})$.
Now, let $U' = V(\varepsilon_0, M)$ and $N' = \text{Cl } U'$.
Then

$$\tilde{\Lambda} \times N' \subset N_{1,\tilde{\Lambda}}$$

by (3) of Lemma 5.3
Applying Lemma 5.4 with U, N, a_0, b_0, M being replaced, respectively, by

$U', N', \varepsilon_0, M, 2M$, we obtain an $\varepsilon_0' = \varepsilon_0'(2M) < \min(\varepsilon_0, \varepsilon_1)$ such that (3) of Lemma 5.4 holds for $\varepsilon \leq \varepsilon_0'$ and neighborhoods

$$W_0(\varepsilon, 2M) \quad \text{of } \lambda_0 .$$

Set

$$W = W_0(\varepsilon_0', 2M) \cap W_2 \tag{8}$$

and for any compact $\tilde{\Lambda} \subset W$ and $i=1,2$, let

$$N'_{i,\tilde{\Lambda}} = N_{i,\tilde{\Lambda}}(\varepsilon_0', 2M). \tag{9}$$

Then $\langle N'_{1,\tilde{\Lambda}}, N'_{2,\tilde{\Lambda}} \rangle$ is an index pair in $\tilde{\Lambda} \times N'$, rel. to $\pi(\tilde{\Lambda})$.
Now let $\tilde{\Lambda}$ be an arbitrary compact set in W. Then, clearly, property (2) of Theorem 5.2 holds.
Moreover,

$$N' \subset \operatorname{Int} N \subset \operatorname{Int} \tilde{N} .$$

Set $N_1' := N'$, $N_1 := N$,

$$N_2' = \tilde{\Lambda} \times (\{x \in \tilde{U} \mid t^+_{\lambda_0}(x) \leq 3M\} \cap N') \tag{10}$$

$$N_2 = \tilde{\Lambda} \times (\{x \in \tilde{U} \mid t^+_{\lambda_0}(x) \leq 5M\} \cap N) . \tag{11}$$

As in the proof of Theorem I.12.3, it is easily seen that $\langle N_1', N_2' \rangle$ and $\langle N_1, N_2 \rangle$ are index pairs in N' and N, respectively, relative to π_{λ_0}.
Now the inclusions in (3) of Theorem 5.2 are proved using (7), (10), (11) and Lemma 5.5.
The proof is complete.
In order to apply Theorem 5.2 to the categorial Morse index, we need to know that the sets $\tilde{\Lambda} \times N'$ and $\tilde{\Lambda} \times N$ are strongly $\pi(\tilde{\Lambda})$-admissible. It seems that this requires an extra assumption on the map $\alpha: \Lambda \to S$.
By Proposition 5.1 it is enough to assume that for every sequence $\{\lambda_n\}$ in $\tilde{\Lambda}$, N' (resp. N) is strongly $\{\pi_{\lambda_0}\}$-admissible. This motivates the following concept.

Definition 5.6

A map $\alpha:\Lambda\to S$ is called strictly S-continuous, if α is S-continuous and for every λ_0 there is a neighborhood $\tilde{W}$ of λ_0 and a closed set $\tilde{N}$ in X such that whenever $\tilde{\Lambda}\subset\tilde{W}$ is compact and $\{\lambda_n\}$ is a sequence in $\tilde{\Lambda}$, then $\tilde{N}$ is $\{\pi_{\lambda_n}\}$-admissible.

Remarks:

(1) Notice that by the definition of S-continuity, we may also assume that $\tilde{N}$ is strongly π_λ-admissible for all $\lambda\in\tilde{\Lambda}$.
(2) The difference between S-continuity and strict S-continuity is that the sequence $\{\lambda_n\}\subset\tilde{\Lambda}$ may converge to some $\mu\in\tilde{\Lambda}$, $\mu\neq\lambda_0$.
(3) In all applications of the index to differential equations in Chapter II, all S-continuous maps $\alpha:\Lambda\to S$ considered, are, in fact, strictly S-continuous.
In order to describe precisely the continuation of the categorial Morse index along paths, we need an abstract concept:

Let K be an arbitrary category and C, C' be subcategories of K which are connected simple systems. Let $M=M(C,C')$ be the set of all morphisms $f:A\to A'$ in K with $A\in \mathrm{Obj}(C)$, $A'\in \mathrm{Obj}(C')$. Elements $f,g\in M$, $f:A\to A'$, $g:B\to B'$ are called equivalent ($f\sim g$) if

$$f = \psi\circ g\circ\varphi$$

where $\varphi:A\to B$ and $\psi:B'\to A'$ are the unique morphisms in C and C' respectively.
$\sim$ is easily seen to be an equivalence relation on M. The equivalence class $F=[f]$ of $f:A\to A'$ is called a morphism from C to C'. We write $F:C\to C'$.
If $F:C\to C'$ is a morphism, C is an object of C and C' is an object of C', then clearly there is a unique morphism $h:C\to C'$ in K such that $[h]=F$.
Thus, if $F:C\to C'$ and $F:C'\to C''$ are two morphisms, then we may choose morphisms $h:C\to C'$ and $h:C'\to C''$ in K such that $F=[h]$ and $F'=[h']$.
Define $F'\circ F=[h'\circ h]$.
It is easily seen that this definition is independent of the choice of the representatives h and h'.
If $F:C\to C'$ contains an isomorphism $f:A\to A'$ then every element of F is an isomorphism. We can then define $F^{-1}=[f^{-1}]$.
Finally, we define the identity morphism $I_C:C\to C$ to be $I=[I_C]$, where $I_C:C\to C$ is the identity morphism of C. It is clear that I_C is the set of all morphisms in C.

Using these definitions, the following result is clear:

Proposition 5.7

If K is an arbitrary category, then there is a category L whose objects are small subcategories of K which are connected simple systems and the morphisms $F:C\to C'$ are defined as above.

We also need the following simple result:

Proposition 5.8

Let A be a topological space and $B\subset A$ be closed. Let Λ be a topological space which is contractible to a point $\lambda_0\in\Lambda$. Then the pointed spaces $((A\times\Lambda)/(B\times\Lambda), [B\times\Lambda])$ and $(A/B,[B])$ are of the same homotopy type. The homotopy equivalence is given by the maps f and g defined below.

Proof:

Let $q:A\times\Lambda\to(A\times\Lambda)/(B\times\Lambda)$ and $q':A\to A/B$ be the quotient maps.
Define $F:A\times\Lambda\to A/B$ as $F(a,\lambda)=q'(a)$ and $G:A\to(A\times\Lambda)/(B\times\Lambda)$ as $G(a)=q(a,\lambda_0)$.
Then F (resp. G) induces a continuous base-point preserving map $f:(A\times\Lambda)/(B\times\Lambda)\to A/B$ (resp. $g:A/B\to(A\times\Lambda)/(B\times\Lambda)$) such that

$$F = f\circ q \quad (\text{resp. } G = g\circ q') .$$

Since Λ is contractible to λ_0, there is a continuous map $h:\Lambda\times[0,1]\to\Lambda$ with $h(\lambda,0)\equiv\lambda$ and $h(\lambda,1)\equiv\lambda_0$ for $\lambda\in\Lambda$.
Define $H:(A\times\Lambda)\times[0,1]\to A\times\Lambda$ by

$$H(a,\lambda,t) = (a,h(\lambda,t)).$$

H induces a base-point preserving continuous map
$\overline{H}:(A\times\Lambda)/(B\times\Lambda)\times[0,1]\to(A\times\Lambda)/(B\times\Lambda)$ such that $\overline{H}(q(a,\lambda),t)=qH(a,\lambda,t)$.
Now, clearly, $f\circ g=\mathrm{Id}_{A/B}$ and $g\circ f=\overline{H}(\cdot,1)$, $\overline{H}(\cdot,0)=\mathrm{Id}_{(A\times\Lambda)/(B\times\Lambda)}$.
This completes the proof of the proposition.

Now let $\alpha:\Lambda\to S$ be S-continuous. Suppose that $K(\tilde{\Lambda})$ has a strongly $\pi(\tilde{\Lambda})$-admissible isolating neighborhood $N_{\tilde{\Lambda}}\subset\tilde{\Lambda}\times X$ relative to $\pi(\tilde{\Lambda})$.
Let $\langle N_{1,\tilde{\Lambda}}, N_{2,\tilde{\Lambda}}\rangle$ be an index pair in $N_{\tilde{\Lambda}}$, relative to $\pi(\tilde{\Lambda})$. Fix $\lambda\in\tilde{\Lambda}$ and set

$N_\lambda=\{x\in X \mid (\lambda,x)\in N_{\tilde{\Lambda}}\}$
$N_{i,\lambda}=\{x\in X \mid (\lambda,x)\in N_{i,\tilde{\Lambda}}\}$
for $i=1,2$.

Then it is easily checked that N_λ is a strongly π_λ-admissible isolating neighborhood of K_λ, relative to π_λ, and $\langle N_{1,\lambda}, N_{2,\lambda}\rangle$ is an index pair in N_λ.
There is a continuous base point-preserving map
$\varphi: N_{1,\lambda}/N_{2,\lambda} \to N_{1,\tilde{\Lambda}}/N_{2,\tilde{\Lambda}}$ such that $\varphi([x]) = [(\lambda,x)]$.
φ is the composition of the natural homeomorphism
$\varphi_1: N_{1,\lambda}/N_{2,\lambda} \to (\{\lambda\}\times N_{1,\lambda})/(\{\lambda\}\times N_{2,\lambda})$ and the inclusion induced map
$\varphi_2: (\{\lambda\}\times N_{1,\lambda})/(\{\lambda\}\times N_{2,\lambda}) \to N_{1,\tilde{\Lambda}}/N_{2,\tilde{\Lambda}}$.
By abuse of terminology, we will call φ <u>inclusion induced</u>.
Let f be the homotopy class of φ. Then f is a morphism in $\mathcal{K}$:=HT* from $(N_{1,\lambda}/N_{2,\lambda},[N_{2,\lambda}])$ to $(N_{1,\tilde{\Lambda}}/N_{2,\tilde{\Lambda}},[N_{2,\tilde{\Lambda}}])$.
By the preceding results, f induces a morphism $F=F(\alpha;\lambda,\tilde{\Lambda}):=[f]$

$$F: I(\pi_\lambda,K_\lambda) \to I(\pi(\tilde{\Lambda}),\ K(\tilde{\Lambda})) \quad . \tag{12}$$

Recall that $I(\pi,K)$ is the categorial Morse index of the pair (π,K).
F is called <u>inclusion induced</u>.

We now obtain the following result:

<u>Theorem 5.9</u>

<u>Let $\alpha:\Lambda\to\mathcal{S}$ be strictly $\mathcal{S}$-continuous. Then for every $\lambda_0\in\Lambda$ there is a neighborhood W of λ_0 in Λ such that for every compact $\tilde{\Lambda}\subset W$ and every $\lambda\in\Lambda$ the inclusion induced morphism F in (12) is defined.</u>
<u>If, in addition, $\tilde{\Lambda}$ is contractible to a point $\lambda\in\tilde{\Lambda}$, then F is an isomorphism (i.e. F^{-1} exists).</u>

<u>Proof</u>:

By strict $\mathcal{S}$-continuity, there is a closed set $\tilde{N}$ and a neighborhood $\tilde{W}$ of λ_0 in Λ such that for every $\lambda\in\tilde{W}$, $\tilde{N}$ is a strongly π_λ-admissible isolating neighborhood of K_λ, relative to π_λ, and $\tilde{N}$ is $\{\pi_{\lambda_n}\}$-admissible for every compact set $\tilde{\Lambda}\subset W$ and every sequence $\{\lambda_n\}\subset\tilde{\Lambda}$. Let $W\subset\tilde{W}$ be a neighborhood of λ_0 for which all conclusions of Theorem 5.2 hold. Let $\tilde{\Lambda}\subset W$ be compact and $\lambda\in\tilde{\Lambda}$. By Theorem 5.2, $N\subset\tilde{N}$ and $\tilde{\Lambda}\times N$ is an isolating neighborhood of $K(\tilde{\Lambda})$, relative to $\pi(\tilde{\Lambda})$. By Proposition 5.1, $\tilde{\Lambda}\times N$ is strongly $\pi(\tilde{\Lambda})$-admissible, and so the morphism F in (12) is well-defined.
Now assume that $\tilde{\Lambda}$ is contractible to a point $\mu\in\tilde{\Lambda}$.
Use Theorem 5.2 and consider the following sequence of continuous inclusion induced maps

$$N'_{1,\tilde{\Lambda}}/N'_{2,\tilde{\Lambda}} \overset{b_1}{\to} (\tilde{\Lambda}\times N'_1)/(\tilde{\Lambda}\times N'_2) \overset{b_2}{\to} N_{1,\tilde{\Lambda}}/N_{2,\tilde{\Lambda}} \overset{b_3}{\to} (\tilde{\Lambda}\times N_1)/(\tilde{\Lambda}\times N_2) \ .$$

We claim that $b_3 o b_2$ is a homotopy equivalence. Assume this claim for a moment. By Theorem I.9.4, the morphism $b_2 o b_1$ is homotopy equivalence. Now Lemma I.12.4 shows that b_i, i=1,2,3, are homotopy equivalences. Now consider the sequence of inclusion induced maps

$$N'_{1,\lambda}/N'_{2,\lambda} \overset{c_1}{\to} N'_1/N'_2 \overset{c_2}{\to} N_{1,\lambda}/N_{2,\lambda} \overset{c_3}{\to} N_1/N_2$$

Here $\quad N'_{i,\lambda} = \{x \mid (\lambda,x) \in N'_{i,\tilde{\Lambda}}\}$

$$N_{i,\lambda} = \{x \mid (\lambda,x) \in N_{i,\tilde{\Lambda}}\}$$

for i=1,2.

By our previous considerations $\langle N'_{1,\lambda}, N'_{2,\lambda}\rangle$ and $\langle N_{1,\lambda}, N_{2,\lambda}\rangle$ are index pairs in N' (resp. N) relative to π_λ. Therefore, by Theorem I.9.4, $c_2 o c_1$ and $c_3 o c_2$ are homotopy equivalences. Thus Lemma I.12.4 again implies that c_i, i=1,2,3 are homotopy equivalences.

Consider the sequence

$$N_{1,\lambda}/N_{2,\lambda} \overset{\varphi}{\to} N_{1,\tilde{\Lambda}}/N_{2,\tilde{\Lambda}} \overset{b_3}{\to} (\tilde{\Lambda}\times N_1)/(\tilde{\Lambda}\times N_2) \overset{f}{\to} N_1/N_2 \ .$$

Here φ is the "inclusion" induced map considered before, and f is as in the proof of Proposition 5.8.

Clearly, $f o b_3 o \varphi = c_3$.

Since b_3, c_3 and f are homotopy equivalences, so is φ. This proves that $[\varphi]$ is an isomorphism in HT* so F is an isomorphism.

The theorem is proved except for the claim that $b_3 o b_2$ is a homotopy equivalence.

Consider the sequences

$$N'_1/N'_2 \overset{a_1}{\to} N_1/N_2 \overset{g_1}{\to} (\tilde{\Lambda}\times N_1)/(\tilde{\Lambda}\times N_2)$$

$$N'_1/N'_2 \overset{g_2}{\to} (\tilde{\Lambda}\times N'_1)/(\tilde{\Lambda}\times N'_2) \overset{a_2}{\to} (\tilde{\Lambda}\times N_1)/(\tilde{\Lambda}\times N_2) \ .$$

Here a_1, a_2 are inclusion induced and g_1, g_2 are as in the proof of Proposition 5.8. Clearly, $g_1 o a_1 = g_2 o a_2$.

By Proposition 5.8, g_1 and g_2 are homotopy equivalences, and by Theorem I.9.4 a_1 is a homotopy equivalence. Thus a_2 is a homotopy equi-

valence. However, $a_2=b_3 o b_2$ and the claim follows.
We are now in a position to define the continuation of the categorial Morse index along paths.
Let ρ be a topological space ρ is called an <u>arc</u> if ρ is a homeomorphic image of the interval $[0,1]$, i.e., if there exists a homeomorphism $\tau:[0,1]\to\rho$. The points $\tau(0)$ and $\tau(1)$ are called the <u>endpoints</u> of $\rho(\{\tau(0),\tau(1)\}$ is obviously independent of the choice of $\tau)$. Fix an ordered pair (b_1,b_2) of endpoints of ρ, and let T be the class of all homeomorphisms $\tilde{\tau}:[0,1]\to\rho$ such that $\tilde{\tau}(0)=b_1$, $\tilde{\tau}(1)=b_2$.

Fix $\tau\in T$. Define the relation $\leq$ on ρ by

$$b \leq b' \text{ if and only if } \tilde{\tau}^{-1}(b) \leq \tilde{\tau}^{-1}(b').$$

This obviously is an order relation on ρ and $\leq$ is independent of $\tilde{\tau}\in T$.
Let $[b,b']=\{a\in\rho \mid b\leq a\leq b'\}$.

Now suppose that $\alpha:\Lambda\to S$ is strictly S-continuous and let $\rho\subset\Lambda$ be an oriented arc.
Using our preceding results we see that there is a finite open covering $\{W_l \mid l=1,\dots,p\}$ of ρ in Λ such that for every $l=1,\dots,p$, and every compact set $\tilde{\Lambda}\subset W_l$ which is contractible to a point the following properties hold:

(1) $(\pi(\tilde{\Lambda}),K(\tilde{\Lambda}))\in S(\tilde{\Lambda}\times X)$.
(2) For every $\lambda\in\tilde{\Lambda}$ the inclusion induced morphism $F(\alpha;\lambda,\tilde{\Lambda})$ defined in (12) is an isomorphism.
Let $P=(\lambda_0,\dots,\lambda_n)$ be a partition of ρ subordinate to the above covering; in other words, $\lambda_i\in\rho$, $b_1=\lambda_0<\lambda_1<\dots<\lambda_n=b_2$, and for every $i=0,\dots,n-1$, there is an $l=1,\dots,p$ with $[\lambda_i,\lambda_{i+1}]\subset W_l$. Let
$F(\alpha,\rho,(b_1,b_2),P)=f_{n-1}\circ\dots\circ f_1\circ f_0$, where
$f_i=F^{-1}(\alpha,\lambda_{i+1},[\lambda_i,\lambda_{i+1}])\circ F(\alpha,\lambda_i,[\lambda_i,\lambda_{i+1}])$, $\quad i=0,\dots,n-1$

<u>Proposition 5.10</u>

$F(\alpha,\rho,(b_1,b_2),P)$ <u>is independent of the choice of</u> P.

<u>Proof</u>:

Let $P=(\lambda_0,\dots,\lambda_n)$ be a partition subordinate to the above covering.
Let $j\in\{0,\dots,n-1\}$ and let $\lambda^*\in\rho$ be such that $\lambda_j<\lambda^*<\lambda_{j+1}$. Consider

$$P = (\lambda_0,\dots,\lambda_j,\lambda^*,\lambda_{j+1},\dots,\lambda_n) .$$

Then $$F(\alpha,\rho,(b_1,b_2),P^*) = f_n\circ\ldots\circ f_{j+1}\circ h\circ g\circ f_{j-1}\circ\ldots\circ f_0 ,$$

where f_i, $i=0,\ldots,n-1$ is as above (with respect to P), and

$$h = F^{-1}(\alpha,\lambda_{j+1},[\lambda^*,\lambda_{j+1}])^{-1}\circ F(\alpha,\lambda^*,[\lambda^*,\lambda_{j+1}]),$$

$$g = F^{-1}(\alpha,\lambda^*,[\lambda_j,\lambda^*])\circ F(\alpha,\lambda_j,[\lambda_j,\lambda^*]) .$$

Using Theorem 5.9 we easily prove that

$$g = F^{-1}(\alpha,\lambda^*,[\lambda_j,\lambda_{j+1}])\circ F(\alpha,\lambda_j,[\lambda_j,\lambda_{j+1}]) ,$$

$$h = F^{-1}(\alpha,\lambda_{j+1},[\lambda_j,\lambda_{j+1}])\circ F(\alpha,\lambda^*,[\lambda_j,\lambda_{j+1}]) ,$$

and hence $h\circ g=f_j$, which implies

$$F(\alpha,\rho,(b_1,b_2),P) = F(\alpha,\rho,(b_1,b_2),P^*) .$$

This proves the proposition (by induction).
Thus we may define a morphism (in HT*)

$$F(\alpha,\rho,(b_1,b_2)) : I(\pi_{b_1},K_{b_1}) \to I(\pi_{b_2},K_{b_2}) ,$$

by

$$F(\alpha,\rho,(b_1,b_2)) = F(\alpha,\rho,(b_1,b_2),P) ,$$

where $P=(\lambda_0,\ldots,\lambda_n)$ is any partition of ρ such that $\lambda_0=b_1$ and $\lambda_n=b_2$, subordinate to some covering $\mathcal{W}$ satisfying the conditions stated above. $F(\alpha,\rho,(b_1,b_2))$ is called the <u>continuation</u> of $I(\pi_{b_1},K_{b_1})$ along the oriented arc $\langle\rho,(b_1,b_2)\rangle$ in α. If Λ is an arc, then α is called a <u>path in</u> S. If α is a path in S and $\rho=\Lambda$, then we call $F(\alpha,\rho,(b_1,b_2))$ the <u>continuation of</u> $I(\pi_{b_1},K_{b_1})$ <u>along the oriented path</u> $\langle\alpha,(b_1,b_2)\rangle$. We then write $F(\alpha,b_1,b_2):=F(\alpha,\rho,(b_1,b_2))$.

<u>Definition 5.11</u>

Two strictly S-continuous paths $\alpha:[0,1]\to S$ and $\bar\alpha:[0,1]\to S$ with $\alpha(0)=\bar\alpha(0)$ and $\alpha(1)=\bar\alpha(1)$ are called <u>strictly</u> S-<u>homotopic</u> if there is a strictly S-continuous mapping $H:[0,1]\times[0,1]\to S$ such that for all $\lambda\in[0,1]$, $s\in[0,1]$, $H(\lambda,0)=\alpha(\lambda)$, $H(\lambda,1)=\bar\alpha(\lambda)$, $H(0,s)=\alpha(0)$, $H(1,s)=\alpha(1)$. H is called a strict S-homotopy from α to $\bar\alpha$.

Now we can state the following

Theorem 5.12

Let $\alpha:[0,1]\to S$ and $\overline{\alpha}:[0,1]\to S$ be strictly S-homotopic. Then

$$F(\alpha,0,1) = F(\overline{\alpha},0,1) .$$

Proof:

Let $P=(\lambda_0,\dots,\lambda_n)$ and $P^*=(s_0,\dots,s_m)$ be two partitions of $[0,1]$, $s_0=\lambda_0=0$, $\lambda_n=s_m=1$. If $0\leq i\leq n-1$ and $0\leq j\leq m-1$ are integers, let $Q_{ij}=[\lambda_i,\lambda_{i+1}]\times[s_j,s_{j+1}]$. Also for $i=0,\dots,n$ and $j=0,\dots,m$, set $q_i^j=(\lambda_i,s_j)\in[0,1]\times[0,1]$. Let $\rho=([0,1]\times\{0\})$, $\overline{\rho}=([0,1]\times\{1\})$. By our assumptions and Proposition 5.13 below,

$$F(\alpha,0,1) = F(H,\rho,(q_0^0,q_n^0)) ,$$

and

$$F(\overline{\alpha},0,1) = F(H,\overline{\rho},(q_0^m,q_n^m)) .$$

Hence, to prove the theorem, it is sufficient (by a simple diagram chase) to prove that for every $i=0,\dots,n-1$, $j=0,\dots,m-1$, $A_{ij}=B_{ij}$, where

$$A_{ij}=F(H,([\lambda_i,\lambda_{i+1}]\times\{s_j\})\cup(\{\lambda_{i+1}\}\times[s_j,s_{j+1}]),\ (q_i^j,q_{i+1}^{j+1})) ,$$

$$B_{ij}=F(H,(\{\lambda_j\}\times[s_j,s_{j+1}])\cup([\lambda_i,\lambda_{i+1}]\times\{s_{j+1}\}),\ (q_i^j,q_{i+1}^{j+1})) .$$

We can choose P and P* in such a way that whenever $W\subset Q_{ij}$ is compact and contractible to a point and $(\lambda,s)\in W$, then the inclusion induced morphism $F(H;(\lambda,s),W)$ is an isomorphism.

It follows that

$$A_{ij}=(F^{-1}(H;q_{i+1}^{j+1},\{\lambda_{i+1}\}\times[s_j,s_{j+1}])\circ F(H;q_{i+1}^{j},\{\lambda_{i+1}\}\times[s_j,s_{j+1}]))$$

$$\circ(F^{-1}(H;q_{i+1}^{j},[\lambda_i,\lambda_{i+1}]\times\{s_j\})\circ F(H,q_i^j,[\lambda_i,\lambda_{i+1}]\times\{s_j\}))$$

$$= (F^{-1}(H;q_{i+1}^{j+1},Q_{ij})\circ F(H;q_{i+1}^{j},Q_{ij}))\circ(F^{-1}(H;q_{i+1}^{j},Q_{ij})\circ F(H;q_i^j,Q_{ij}))$$

$$= F^{-1}(H;q_{i+1}^{j+1},Q_{ij})\circ F(H;q_i^j,Q_{ij}) .$$

Doing the same for B_{ij} we conclude immediately that $A_{ij}=B_{ij}$. The theorem is proved.

Proposition 5.13

Let $\alpha:\Lambda\to S$ be strictly S-continuous and let $\rho\subset\Lambda$ be an oriented arc in Λ with initial point b_1 and end point b_2. If $\tau:[0,1]\to\rho$ is a homeomorphism such that $\tau(0)=b_1$ and $\tau(1)=b_2$, then
$F(\alpha,\rho,(b_1,b_2))=F(\alpha\circ\tau, 0,1)$.

The proof is left to the reader as a simple exercise.

Bibliographical Notes and Comments

Introduction:

For an exposition of the Conley index theory for flows on compact spaces see Conley [1], Conley and Zehnder [1] (cf. also Smoller [1] and Salamon [1]).

For an introduction to the index theory on noncompact spaces, see Rybakowski [11], and also Mawhin and Rybakowski [1].

Sections 1.1, 1.2, 1.3:

For more information on local semiflows (local semidynamical systems) see Bhatia and Hajek [1]. Notice, however, that their terminology and notation sometimes differs from ours. Thus, e.g. we do not define $A\pi S$ unless $S\times A\subset D$.
Also, our definition of negative invariance is only slightly stronger than their "weak negative invariance".
Isolated invariant sets and isolating block were first defined in Conley and Easton [1] and subsequently in Churchill [1] and Wilson and Yorke [1]. All those definitions are given for a (global) two-sided flow on a compact manifold or, more generally, on compact metric spaces.
The definitions 3.1, 3.2 and 3.3 first appeared in Rybakowski [4]. They retain the useful properties of definitions given by the previously mentioned authors, but also take into account the complications due to the fact that π is only a local (one-sided) semiflow on a general not necessarily compact metric space X.
Theorem 3.4 appeared in Dunbar, Rybakowski and Schmitt [1], extending and simplifying a previous result from Freedman and Waltman [1] and Butler and Waltman [1]. It can be used to prove persistence in various population models.
Lemma 3.8 and Theorem 3.7 are essentially contained in Rybakowski and Zehnder [1].
For more information on evolution equations see Henry [1], Pazy [1], Tanabe [1] and Segal [1].

Section 1.4:

Admissibility was defined in Rybakowski [4]. For Theorems 4.3, 4.4 cf. also Rybakowski [7], [9].
Theorem 4.5 collects various results from Rybakowski [4].

Section 1.5:

Theorem 5.1 is contained in Rybakowski [4], Corollary 5.5 is implicitly contained in the proof of Theorem 3.4 in Rybakowski [7].

Section 1.6:

For more information on homotopies see Spanier [1], Switzer [1], and Hilton [1].
Proposition 6.2 was proved in Kurland [1].

Sections 1.7 and 1.8:

For flows on compact spaces, index pairs were first defined in Conley [1].
Definition 7.1 appeared in Rybakowski [4] and is just Conley's definition except what we require N_i, i=1,2 to be only closed but not necessarily compact.
Quasi-index pairs were first defined in Rybakowski [6] (see also Rybakowski [5]). Proposition 7.3 was stated without proof in Rybakowski [6].
Propositions 8.1 and 8.2 were also stated in Rybakowski [6] without proof. The proofs of some partial results were given in Rybakowski [4]. Rather complete proofs of the corresponding results for the easier case of flows on compact spaces were given by Kurland [1].

Sections 1.9 and 1.10:

Theorem 9.1 was proved in Rybakowski [6]. It extends earlier results of Conley [1] (with proofs in Kurland [1]).
Theorem 10.1 also appeared with a somewhat different proof in Rybakowski [4].
Theorems 10.4 and 10.6 are taken from Rybakowski [9]. There are many different applications of the cohomological index not discussed in this book. We only mention Amann and Zehnder [1], Angenent [1], Churchill [1], [2], Conley and Smoller [1], [2], [3], Conley and Zehnder [1], Dancer [1], Easton [1], Montgomery [1], Smoller [1]. For more references see Conley [1] and the concluding remarks of this book.
Using a result of Mañe one can show that for large classes of infinite-dimensional semiflows, only a finite number of the Cech-cohomology group of the homotopy index can be nontrivial (see Rybakowski [4]). Incidentally this can be viewed as a heuristic explanation of why the so-called Ważewski principle (see Ważewski [1], Conley [1] cf. also Rybakowski [1], [2], [3]) "works" in infinite-dimensions, des-

pite the fact that the infinite-dimensional unit sphere is a strong deformation retract of the closed unit ball.

Section 1.11:

Theorem 11.1 appeared in Rybakowski [7] and, in a special case, in Rybakowski [4]. Irreducibility was defined in Rybakowski [8]. The latter work also contains Theorem 11.5 (with a slightly incorrect proof) and Theorem 11.6.
Theorem 11.8 and Corollary 11.9 are (implicitly) contained in Rybakowski [7].

Section 1.12:

The results of this section are taken from Rybakowski [4]. The definition of $\mathcal{S}$-continuity was first given in Rybakowski [6].

Section 2.1:

For more information on elliptic equations see Friedman [1] and Gilbarg and Trudinger [1].

Sections 2.3 and 2.4:

The index product-formula for a flow π generated by a finite-dimensional ODE first appeared in Amann and Zehnder [1]. Its proof, due to Conley, is a simple application of Palmer generalization of the Hartman-Grobman theorem (Palmer [1]). This latter result is no longer true for (one-sided) semiflows. This explains why Theorem 3.1 (which is taken from Rybakowski [9]) has such a long proof.
The results in section 2.4 are taken from Rybakowski [10].

Sections 2.5 and 2.6:

Most of the results in these sections are taken from Rybakowski [10]. Some related results for a damped wave equation appeared in Rybakowski [7].
Theorem 5.6 is essentially the bifurcation theorem of Krasnoselskii and Rabinowitz (see Krasnoselskii [1] p. 332 and Rabinowitz [1]).
Corollary 5.8 extends earlier results of Amann and Zehnder [1] obtained under the more stringent assumption that $\frac{\partial f}{\partial s}(x,s)$ exists globally, is continuous and bounded on $\overline{\Omega}\times\mathbb{R}$.
These assumptions enable the authors to apply a finite-dimensional reduction procedure due to Amann [2], thus being able to use Conley index in finite-dimensions. Proposition 6.4 appeared in Conley and Zehnder [1].

Section 2.7:

The results of this section are refinements of the corresponding results in Rybakowski [7] and [10].
The existence of positive solutions of (24) can also be obtained using the topological degree (see Peitgen and Schmitt [1]).

Section 2.8:

The results in this section are essentially taken from Rybakowski [12]. They are motivated by the coincidence degree method developed by Mawhin (Mawhin [1], [2], [3]). The example in Proposition 8.6 is a modification of an earlier example given in Rybakowski [3].

Sections 3.1 and 3.2:

The results of this section are essentially contained in Rybakowski and Zehnder [1] extending earlier results of Conley [1]. The term "block pair" appears here for the first time.
Morse decompositions for a class of delay equations are studied in Mallet-Paret [1].

Section 3.3:

Theorem 3.5 is an improvement of a corresponding result from Rybakowski and Zehnder [1], where an additional finite-rank assumption is made. It extends earlier results of Conley and Zehnder [1] to arbitrary homology and cohomology groups and to (one-sided) semiflows on not necessarily compact metric space satisfying the admissibility assumption.
Theorem 3.8 is contained in Rybakowski [13] and was obtained independently from an earlier similar result of Dancer [1], who proves a corresponding formula for an invariant set consisting of one (critical) point. The latter work contains a number of very interesting results on the cohomological index of a degenerate critical point for a gradient ODE as well as some applicationd to PDEs in the spirit of Rybakowski [7], [10]. For some applications of the topological degree in Morse theory see also Hofer [1], [2].

Section 3.4:

For Theorem 4.2 and Corollary 4.3, cf. Rybakowski [15]. Theorem 4.4 is essentially contained in Rybakowski [7]. For the remaining results of this section, see Rybakowski [14]. The concept of critical groups is due to Rothe [1]. It has been used by Chang [1] to prove some results of Amann and Zehnder [1] by using the classical Morse theory (instead of the Conley index) generalized to the case of degenerate critical points.

Section 3.5:

Theorem 5.2 was stated (without proof) in Rybakowski [6]. The continuation results in the latter work corresponding to Theorem 5.12 are incorrectly derived under the sole assumption of S-continuity. However, it seems that strict S-continuity is what is needed here. For flows on compact spaces, Theorem 5.12 is due to Kurland [2], [3], although a related result was earlier proved by Montgomery [2].

Concluding remarks

Several important aspects of the homotopy index theory were not discussed in this book. Thus we did not consider the so-called connection index and the connection matrix. The interested reader is referred to Conley [1], Franzosa [1], Franzosa and Mischaikow [1], Kurland [2], [3], [4] and Rybakowski [6]. (See also Mischaikow [1] for some applications).

Suppose that π is a local semiflow on X which is equivariant with respect to some group action on X. Then it is not difficult to see (by inspecting the construction presented in this book) that there exist symmetric isolating blocks and symmetric index pairs. This leads to equivariant versions of the homotopy and Morse index.
In the compact case, the equivariant Conley index was discussed in Floer [1], [2], cf. also Pacella [1], Benci and Pacella [1].

For a definition of the Conley-type index for maps, see the recent paper of Robbin and Salamon [1]. This should have interesting applications to difference equations.

Bibliography

H. Amann:
[1] Fixed point equations and nonlinear eigenvalue problems in ordered Banach spaces, SIAM Review 18 (1976), 620-709.

[2] Saddle points and multiple solutions of differential equations, Math. Z. 169 (1979), 127-166.

H. Amann and E. Zehnder:
[1] Nontrivial solutions for a class of nonresonance problems and applications for nonlinear differential equations, Ann. Scuola Norm. Sup. Pisa, IV, Vol. VII, (1980), 539-603.

S.B. Angenent:
[1] The periodic orbits of an area preserving twist map, preprint, University of Leiden, 1985.

E. Artin and H. Braun:
[1] "An introduction to algebraic topology", Charles E. Merrill Publishing Comp., Columbus, Ohio, 1969.

V. Benci and F. Pacella:
[1] Morse theory for symmetric functional on the sphere and an application to a bifurcation problem, Nonl. Anal. Theory Meth. Appl. 9 (1985), 763-773.

N. Bhatia and O. Hájek:
[1] "Local semi-dynamical systems", Lect. Notes in Math., Vol. 90, Springer Verlag, Berlin, New York, 1969.

G. Butler and P. Waltman:
[1] Persistence in dynamical systems, J. Diff. Eq. 63 (1986), 255-263.

J. Carr:
[1] "Application of centre manifold theory", Appl. Math. Sc. 35, Springer Verlag, New York, Berlin, 1981.

K.C. Chang:
[1] Solutions of asymptotically linear operator equations via Morse theory, Comm. Pure Appl. Math. 34 (1981), 693-712.

S.N. Chow and J.K. Hale:
[1] "Methods of Bifurcation Theory", Springer Verlag, New York, Berlin, 1982.

R. Churchill:
[1] Isolated invariant sets in compact metric spaces, J. Diff. Eq. 12 (1972), 330-352.

[2] Invariant sets which carry cohomology, J. Diff. Eq. 13 (1973), 523-550.

C.C. Conley:
[1] "Isolated invariant sets and the Morse index", CBMS 38, AMS, Providence, 1978.

C.C. Conley and R.W. Easton:
[1] Isolated invariant sets and isolating blocks, TAMS 158 (1971), 35-61.

C.C. Conley and J. Smoller:
[1] and [2] Shock waves as limits of progressive wave solutions of higher-order equations, I and II, Comm. Pure Appl. Math. 24 (1971), 459-472, and 25 (1972), 131-146.

[3] Remarks on traveling wave solutions of non-linear diffusion equations, in: Lect. Notes in Math. 525 (ed. P. Hilton) (1975), 77-89.

C.C. Conley and E. Zehnder:
[1] Morse type index theory for flows and periodic solutions for Hamiltonian systems, Comm. Pure Appl. Math. 37 (1984), 207-253.

E.N. Dancer:
[1] Degenerate critical points, homotopy indices and Morse inequalities, Journal Reine Angew. Math. 350 (1984), 1-22.

K. Deimling:
[1] "Nonlinear Functional Analysis", Springer Verlag, Berlin, New York, 1985.

S. Dunbar, K.P. Rybakowski and K. Schmitt:
[1] Persistence in models of predator-prey populations with diffusion, J. Diff. Eq. 65 (1986), 117-138.

R.W. Easton:
[1] Isolating blocks and symbolic dynamics, J. Diff. Eq. 17 (1975), 96-118.

A. Floer:
[1] A refinement of the Conley index and an application to the stability of hyperbolic invariant sets, preprint, Univ. Bochum, 1985.

[2] Proof of the Arnold conjecture for surfaces and generalizations to certain Kähler manifolds, preprint, Univ. Bochum, 1985.

R.D. Franzosa:
[1] Index filtrations and connection matrices for partially ordered Morse decompositions, Ph. D. Thesis, Univ. of Wisconsin-Madison, 1984.

R.D. Franzosa and K. Mischaikow:
[1] The connection matrix theory for semiflows on (not necessarily locally compact) metric spaces, preprint, Brown University, 1986.

H. Freedman and P. Waltman:
[1] Persistence in models of three interacting predator-prey populations, Math. Biosci. 68 (1984), 213-231.

A. Friedman:
[1] "Partial Differential Equations", Holt, Rinehard and Winston, New York, 1969.

D. Gilbarg and N.S. Trudinger:
[1] "Elliptic partial differential equations of second order", Springer Verlag, New York, 1977.

J.K. Hale:
[1] "Theory of Functional Differential Equations", Springer Verlag, New York, 1977.

J.K. Hale and K.P. Rybakowski:
[1] On a gradient-like integro-differential equation, Proc. Roy. Soc. Edinb., 92A (1982), 77-85.

D. Henry:
[1] "Geometric Theory of Semilinear Parabolic Equations", Lect. Notes in Math., vol. 840, Springer Verlag, Berlin, New York, 1981.

P. Hilton:
[1] "General cohomology theory and K-theory", London Math. Soc. Lect. Notes Series, vol. 1, Cambridge Univ. Press, 1971.

H. Hofer:
[1] Variational and topological methods in partially ordered Hilbert spaces, Math. Ann. 261 (1982), 493-514.

[2] A note on the topological degree at a critical point of mountain-pass type, Proc. AMS 90 (1984), 309-315.

M.A. Krasnoselskii:
[1] "Topological methods in the theory of nonlinear integral equations", Pergamon Press, 1963.

S. Kuen and K.P. Rybakowski:
[1] Boundedness of solutions of a system of integro-differential equations, J. Math. Anal. Appl. 112 (1985), 378-390.

H. Kurland:
[1] The Morse index of an isolated invariant set is a connected simple system, J. Diff. Eq. 42 (1981), 234-259.

[2] Homotopy invariants of repeller-attractor pairs, I. The Puppe sequence of a R-A pair, J. Diff. Eq. 46 (1982), 1-31.

[3] Homotopy invariants of repeller-attractor pairs, II. Continuation of R-A pairs, J. Diff. Eq. 49 (1983), 281-329.

[4] Singularly perturbed systems and the Morse-Conley index, unpublished manuscript.

J. Mallet-Paret:
[1] Morse decompositions and global continuation of periodic solutions for singularly perturbed delay equations, in: J.M. Ball (Ed.) Systems of Nonlinear Partial Differential Equations, D. Reidel Publ. Comp. 1983, 351-365.

C.R.F. Maunder:
[1] "Algebraic topology", Cambridge University Press, 1980.

J. Mawhin:
[1] Degré topologique et solutions périodique des systèmes différentielles non linéaires, Bull. Roy. Soc. Liège 38 (1969), 308-398.

[2] Equivalence theorems for nonlinear operator equations and coincidence degree theory for some mappings in locally convex topological spaces, J. Diff. Eq. 12 (1972), 610-636.

[3] "Topological degree methods in nonlinear boundary value problems", CBMS 40, AMS Providence, 1980.

J. Mawhin and K.P. Rybakowski:
[1] On continuation theorems for semilinear equations in Banach spaces: a survey, to appear in: "Nonlinear functional analysis and fixed point theory", T.M. Rassias, ed., North Holland Publ. Comp.

J. Mawhin and M. Willem:
[1] On the generalized Morse Lemma, Bull. Soc. Math. Belg., Serie B, volume 37, fasc. II (1985), 23-29.

[2] "Critical point theory and Hamiltonian systems, Appl. Math. Sci., Springer Verlag, New York, to appear.

K. Mischaikow:
[1] Classification of traveling wave solutions of reaction-diffusion systems, preprint.

J.T. Montgomery:
[1] Cohomology of isolated invariant sets under perturbations, J. Diff. Eq. 13 (1973), 257-299.

[2] On the homotopy index of Conley, J. Diff. Eq. 32 (1979), 32-40.

R.D. Nussbaum:
[1] The fixed point index for local condensing maps, Ann. Mat. Pura Appl. (4) 89 (1971), 217-258.

F. Pacella:
[1] Morse theory for flows in presence of a symmetry group, preprint, University of Wisconsin, 1984.

K. Palmer:
[1] Linearization near an integral manifold, J. Math. Anal. Appl. 51 (1975), 243-255.

A. Pazy:
[1] "Semigroups of linear operators and applications to partial differential equations", Appl. Math. Sc. 44, Springer Verlag, New York, 1983.

H.O. Peitgen and K. Schmitt:
[1] Global topological perturbations of nonlinear elliptic eigenvalue problems, unpublished manuscript.

M. Protter and H. Weinberger:
[1] "Maximum principles in Differential equations", Prentice-Hall, Inc. Englewood Cliffs, N.J., 1967.

D. Puppe:
[1] Homotopiemengen und ihre induzierten Abbildungen I, Math. Z. 69 (1958), 299-344.

P.H. Rabinowitz:
[1] A bifurcation theorem for potential operators, J. Funct. Anal. 25 (1977), 412-424.

J. Robbin and D. Salamon:
[1] Shape theory and dynamical systems, preprint.

E. Rothe:
[1] Critical point theory in Hilbert spaces under regular boundary conditions, J. Math. Anal. Appl. 36 (1971), 377-431.

K.P. Rybakowski:
[1] Ważewski principle for retarded functional differential equations, J. Diff. Eq. 36 (1980), 117-138.

[2] A topological principle for retarded functional differential equations, J. Diff. Eq. 39 (1981), 131-150.

[3] Some remarks on periodic solutions of Carathéodory RFDEs via Ważewski's principle, Riv. Mat. Univ. Parma (4) 8 (1982), 377-385.

[4] On the homotopy index for infinite-dimensional semiflows, TAMS 269 (1982), 351-382.

[5] On the Morse index for infinite-dimensional semiflows, in: "Dynamical Systems II", Bednarek/Cesari, eds., Academic Press (1982), 633-637.

[6] The Morse index, repeller-attractor pairs and the connection index for semiflows on noncompact spaces, J. Diff. Eq. 47 (1983), 66-98.

[7] Trajectories joining critical points of nonlinear parabolic and hyperbolic partial differential equations, J. Diff. Eq. 51 (1984), 182-212.

[8] Irreducible invariant sets and asymptotically linear functional differential equations, Boll. Unione Mat. Italiana (6) 3-B (1984), 245-271.

[9] An index product-formula for the study of elliptic resonance problems, J. Diff. Eq. 56 (1985), 408-425.

[10] Nontrivial solutions of elliptic boundary value problems with resonance at zero, Annali Mat. Pura ed Appl. (IV), Vol. CXXXIX (1985), 237-278.

[11] An introduction to the homotopy index on noncompact spaces, Appendix to "An introduction to infinite-dimensional dynamical systems-geometric theory" by J. Hale et al., Appl. Math. Sciences, Vol. 47, Springer Verlag New York 1983, 147-191.

[12] A homotopy index continuation method and periodic solutions of second order gradient systems, J. Diff. Eq. 65 (1986), 203-218.

[13] On a relation between the Brouwer degree and the Conley index for gradient flows, Bull. Soc. Math. Belg., serie B, volume 37, fasc. II, (1985), 87-96.

[14] On critical groups and the homotopy index in Morse theory on Hilbert manifolds, Rend. Ist. Mat. Univ. Trieste, to appear.

[15] Some recent results in the homotopy index theory in infinite dimensions, Rend. Ist. Mat. Univ. Trieste, vol. XVIII (1986), 83-93.

K.P. Rybakowski and E. Zehnder:
[1] On a Morse equation in Conley's index theory for semiflows on metric spaces, Ergodic Theory Dyn. Systems 5 (1985), 123-143.

D. Salamon:
[1] Connected simple systems and the Conley index of isolated invariant sets, TAMS, 291 (1985), 1-41.

I. Segal:
[1] Non-linear semi groups, Ann. of Math. 78 (1963), 339-364.

J. Smoller:
[1] "Shock waves and reaction-diffusion equations", Springer Verlag, New York, Berlin 1983.

E.H. Spanier:
[1] "Algebraic topology", McGraw-Hill, New York 1966.

R.M. Switzer:
[1] "Algebraic topology-homotopy and homology", Springer Verlag, New York, 1975.

H. Tanabe:
[1] "Equations of evolution", Pitman Publishing Ltd., London, 1979.

T. Ważewski:
[1] Sur un principe topologique de l'examen de l'allure asymptotique des intégrales des équations différentielles, Ann. Soc. Polon. Math. 20 (1947), 279-313.

F.W. Wilson, Jr. and J.A. Yorke:
[1] Lyapunov functions and isolating blocks, J. Diff. Eq. 13 (1973), 106-123.

Index

admissible inclusion induced map 28
admissible pair 50
($\{\pi_n\}$-, π-) admissible set 13
arc 191
attractor 141

base point 26
block pair 148
bounce-off point 9

Categorial Morse index 41
center manifold 77
center unstable manifold 76
classical solution 74
cofibration 11
cofibration property 149
cohomological index 57
collective ω-limit set 141
conditionally completely continuous 103
Conley index 49
connected simple system 41
connection index 199
connection matrix 199
continuation of $I(\pi,K)$ along an oriented arc (path) 192
S-continuous 65
convergence of semiflows 5
$\{\pi_n\}$ converges to π 5
critical group 168
critical point 163

distributional solution 74
does not explode 5

equilibrium 96
equivalent morphisms 187
equivariant Conley index 199
exit ramp 33
explode 5

formal Poincaré-polynomial 156
fractional power space 4
full solution 6

global attractor 103
global semiflow 2
gradient-like 96

heteroclinic orbit 97
homological index 57
homotopic 26
S-homotopic 192
homotopy 26
homotopy equivalent 26
homotopy index 49
homotopy type 27
hyperbolic 87

inclusion induced map 28
inclusion induced morphism 189
index pair 30
index triple 148
invariant set 7
irreducible 60
isolated invariant set 8
isolating block 9
isolating neighborhood 8

join 51

Liapunov function 96
α-limit set 6
ω-limit set 6, 141
ω^*-limit set 6
linear semiflow 57
local center manifold 77
local center unstable manifold 76

local semiflow 2

morphism 26, 187
Morse decomposition 143
Morse index 41

negatively invariant set 7
Nemitski operator 75
no-blow-up condition 5

Palais-Smale condition 164

parabolic equations 4
Poincaré polynomial 156
point-dissipative 103
pointed space 26
positively invariant set 7
N-positively invariant set 29
product local semiflow 53
product semiflow 53

quasi-index pair 31
quasi-potential 169

reducible 60
repeller 141
repeller-attractor pair 141

saddle-point property 87
sectorial operator 4
semiflow 2
smash product 51
solution 4, 6
stable difference operator 3
stable manifold 7
strict egress point 9
strict ingress point 9
strictly S-continuous 187
strictly S-homotopic 192
strongly ($\{\pi_n\}$-, π-) admissible set 13
strongly elliptic 74

topological pair 26

union of all full bounded orbits 94
unstable manifold 7

Ważewski principle 196
wedge sum 51